# Face Detection and Recognition

## Theory and Practice

T0138844

# Face Detection and Recognition
## Theory and Practice

**Asit Kumar Datta**
University of Calcutta, Kolkata, India

**Madhura Datta**
University of Calcutta, Kolkata, India

**Pradipta Kumar Banerjee**
Future Institute of Engineering and Management,
Kolkata, India

CRC Press
Taylor & Francis Group
Boca Raton  London  New York

CRC Press is an imprint of the
Taylor & Francis Group, an **informa** business

CRC Press
Taylor & Francis Group
6000 Broken Sound Parkway NW, Suite 300
Boca Raton, FL 33487-2742

First issued in paperback 2019

© 2016 by Taylor & Francis Group, LLC
CRC Press is an imprint of Taylor & Francis Group, an Informa business

No claim to original U.S. Government works

ISBN-13: 978-1-4822-2654-6 (hbk)
ISBN-13: 978-0-367-37726-7 (pbk)

**Visit the Taylor & Francis Web site at**
**http://www.taylorandfrancis.com**

**and the CRC Press Web site at**
**http://www.crcpress.com**

*To our Teachers, Peers, Friends, Families, and our Mentor*
*Professor C. A. Murthy of the*
*Indian Statistical Institute, Kolkata*

# Contents

# List of Figures

# List of Tables

# *Preface*

Face detection and recognition represent nonintrusive methods for recognizing people, and these are the biometrics of choice in many security applications. Face detection and recognition are some remarkable and important abilities that we use in our daily lives. The main reason for the interest in developing the computer vision-based automated technologies of automated face recognition arises from the serious concerns for public security in today's networked world, where identity verifications for physical and logical access in many facilities are imperative in daily life. Though the first widely accepted algorithm during the 1970s was the eigenface method, which even today is used as a base for many methods, the real impetus came along with the development of computational power and algorithms related to the use of large databases. Therefore, face detection and recognition are still actively researched areas. Many problems related to unconstrained and real-life and real-time environments are yet to be solved to the level of required robustness and accuracy.

The goal of this book is to provide the reader with a description of some available techniques for automated face detection and recognition and is intended for anyone who plans to work in these areas and also for those who want to become familiar with the state-of-the-art techniques. This book is written with two primary motivations. The first is to compile major approaches, algorithms and technologies available for automated face detection and recognition. The second reason is to provide a reference for students, researchers and practitioners working in the areas of image processing, computer vision, biometrics and security, computer graphics and animation. The materials contained in the book support the quest and need for an advanced tutorial, state-of-the-art survey of current technologies and a comprehensive list of major references. Each chapter focuses on a specific topic or system with an introduction to background information, reviews and also some results on typical systems. The usefulness of the book for students and researchers is enhanced by the inclusion of many programs in easily available software.

It may be noted that the evolution of techniques for face detection, recognition and identification is now merging using different available methods of pattern recognition. Therefore it may not be very prudent to maintain individual identity of separate areas. Instead the identities may be merged and may be termed under the general terminology of face recognition. This book is also an attempt in the direction of unification.

Chapter 1 serves the purpose of introducing the subject of face detection, recognition and identification along with an indication of the direction in which future research may aim using cognitive neurophysiology. In Chapter 2, a general review of the available methods in face detection and recognition is presented. Chapter 3 gives an overview of the most commonly used subspace methods for dimensionality reduction in face image processing. Chapter 4 gives an overview of statistical methods applied to face detection. In Chapter 5, face detection with colour and infrared face images is discussed. In Chapter 6, intelligent methods for face detection, which are particularly dominated by the use of the techniques of artificial neural network are presented. In Chapter 7, another important area of face detection in real-time is discussed. Chapter 8 presents a technique of face detection and recognition using set estimation theory. This technique is not very widely used and this chapter may prove to be a stepping stone to activities in this area. In Chapter 9, another interesting area of face recognition using evolutionary algorithms is discussed. Chapter 10 gives an exhaustive discussion on face recognition in frequency domain. The use of correlation filters has proved to be more robust under certain conditions than spatial domain processing of face images. Chapter 11 shows how subspace techniques can be used in a frequency domain. Exhaustive test results are included in these two chapters. In Chapter 12, methods are discussed for the localization of face landmarks helpful in face recognition. Chapter 13 shows methods of generating synthetic face images using set estimation theory. The techniques may help in developing a database of face images using various artifacts. Chapter 14 gives information on major databases of face images available for testing and training of systems. Also in this chapter information on standard vendor tests is included. The book ends with a conclusion note and a list of references and an index.

Asit K. Datta, Madhura Datta and Pradipta K. Banerjee

# Acknowledgment

Thanks are due to a number of individuals who have assisted us during many phases of preparing the manuscript of the book, particularly Aastha Sharma, commissioning editor (India) of CRC Press, without whose support and constant encouragement, this book would not have been possible. Marsha Pronin, the project coordinator, editorial project development, CRC Press deserves special mention as her help was always available to us. Robin Lloyd-Starkes, Alexander Edwards and Jonathan Pennell have very kindly contributed during the production of the book. It was a privilege to work with their team.

# Chapter 1

## Introduction

## 1.1 Introduction

Systems and techniques of face recognition and detection are a subset of an area related to information security, and information security is concerned with the assurance of confidentiality, integrity and availability of information in all forms. There are many tools and techniques that can support the management of information security; however, one of the important issues is the need to correctly authenticate a person. Traditionally, the use of passwords and a personal identification number (PIN) has been employed to identify an individual, but the disadvantages of such methods are that someone else may use the PIN for unauthorized access or the PIN may be easily forgotten.

Many agencies are now motivated to improve security data systems based on body or behavioral characteristics, often called biometrics [1]. Biometric approaches are concerned with identifying an individual by his unique physical characteristics and biological traits. Given these problems, the development of biometrics approaches such as face recognition, fingerprint, iris/retina and voice recognition proves to be a superior solution for identifying individuals over that of PIN codes. The use of biometric techniques not only uniquely identifies an individual, but also minimizes the risk of someone else using the unauthorized identity. Biometric authentication also supports the facets of identification, authentication and nonrepudiation in information security.

The word biometrics, as is used today, is derived from two ancient Greek words, *bios* meaning life and *metrickos* meaning measure. Classically, biometrics refers to the studies related to the biological sciences and is somewhat simply viewed as biological statistics. In modern terminology biometrics, however, refers to the studies related to authentication which is an act of establishing or confirming something (or someone) as authentic, that is, the claims made by or about the thing are true. Various biometric techniques can be broadly categorized as

1. Physical biometrics, which involves some of the physical measurements and includes the characterization of face, fingerprints, iris scans, hand geometry, etc.

2. Behavioral biometrics, which is usually temporal in nature and involves the measurements of performance of a person during the execution of certain tasks, such as speech, signature, gait, keystroke dynamics, etc., and

3. Chemical biometrics, which involves the measurement of chemical cues such as odor and chemical composition of human perspiration.

---

## 1.2   Biometric identity authentication techniques

European explorer Joao de Barros recorded the first known example of fingerprinting, which is a form of biometric authentication. In China during the fourteenth century, merchants used ink to take children's fingerprints for identification purposes. In 1890, Alphonse Bertillon studied body measurements to help in identifying criminals. The police force used the method, called the Bertillonage method, until the cases of false identification were proved. The Bertillonage method was quickly abandoned in favor of fingerprinting, the techniques of which were revived back into use by Richard Edward Henry of Scotland Yard.

Karl Pearson, an applied mathematician, studied biometric research early in the twentieth century at the University College of London. He made important contributions in the field of biometrics by studying statistical history and correlation with the subject. During the middle of the last century, signature based biometric authentication procedures were developed. The biometric field remained stagnant until the interest of military and security agencies grew and biometric technologies were developed beyond the scope of fingerprint and signature-based authentication.

The prevailing political situations throughout the world compelled society to become more conscious regarding security issues of all types. In today's networked world, the need to maintain the security of information or physical properties has become increasingly important. The use of biometrics with different levels of difficulties is becoming almost a necessity in today's life.

Therefore, many techniques were developed using various body parameters for authentication. However, such efforts have also resulted in controversies in which civil liberty groups expressed concern over privacy with respect to the identity issues. Today, biometric laws and regulations are in the process of review and biometric industry standards are being established.

Nevertheless a biometric system can provide two functions, one of which is verification and the other is authentication. Verification is generally related to the database search, and such a process has to be stringent enough so as to employ both these functionalities simultaneously. Seven factors are identified by Jain [2] determine the suitability of unique physical, behavioral and chemical traits for biometric systems. These are

1. Uniqueness, which should be sufficiently different across individuals in a population.

2. Measurability, which is defined as the possibility of acquiring the biological traits by using devices.

3. Universality of the techniques, which means that the access is universally acceptable.

4. Acceptability, which indicates the willingness of the population to utilize the system.

5. Performance should indicate the accuracy and repeatability under given constraints.

6. Permanence, which indicates that the biometric traits are invariant to a certain degree over a period of time.

7. Circumvention, which indicates that the system is not responsive to fake artifacts and rejects mimicry of behavioural traits.

Recently, a new trend has been observed in biometrics that merges human perception to a database in a brain-machine interface. This approach has been referred to as cognitive biometrics. Cognitive biometrics is based on specific responses of the human brain to stimuli which could be used to trigger a computer database search. Cognitive biometric systems are generally developed using the brain response to odor stimuli, facial perception and mental performance. In the near future, these biometric techniques will provide a better solution to recognition and authentication problems without errors so as to equip society to meet the current threats in the domain of information security.

## 1.3   Face as biometric identity

Commonly used biometrics have many drawbacks. Iris recognition is extremely accurate, but expensive for implementation on a wide scale and is not very accepted by people. Fingerprints are reliable as biometrics and non-intrusive, but not suitable for non-collaborative individuals. At present, face recognition seems to be a good compromise between reliability and social acceptance which balances security and privacy well. Face recognition techniques work in unconstrained acquisition conditions and has the great advantage of being able to work in places with large populations of unaware visitors. Because of these advantages face recognition has become one of the most popular biometric techniques.

Facial images are probably the most common biometric characteristic used by humans to make a personal identification. As such, the detection and recognition of faces are the fundamental cognitive abilities that form a basis for our social interactions. From birth, humans experience and participate in face-to-face interactions that contribute to the capability of recognizing faces. Approaches to face recognition are typically based on location and shape of facial attributes, such as the eyes, eyebrows, nose, lips, and chin shape and their spatial relationships.

Faces are complex objects; therefore, detecting and recognizing them are challenging task, despite the relative ease with which humans are able to do. Given an arbitrary image, the goal of face detection and recognition is to determine whether or not there are any faces in an image and, if present, determine the location and extent of each face to be found; then, the face needs to be identified. While this appears to be a trivial task for human beings, it is a very challenging task for any hardware system and therefore has been one of the major research topics in machine vision technology during the past few decades. The efforts of machine detection and recognition of faces can be now combined into a general terminology referred to as *automated face recognition* (AFR). The ultimate goal, however, is to mimic the activities of the human brain performing the tasks of face detection and recognition. The key issue is to understand how it is possible to create representation of faces that achieves the kind of robust face recognition capability which people show in day to day interaction.

Therefore, to replicate the human capability of detection and recognition of faces using machines, the area has drawn the attention of researchers of many hues and specializations, particularly from the fields in image processing, physiology, psychology and computational technology.

## 1.3.1 Automated face recognition system

The first automated face recognition system was developed by Takeo Kanade and was reported in his PhD thesis work in 1973 [3]. It turns out during the initial period of developments, that the techniques of face recognition had the merit of a physiological approach without being intrusive. However, in those days, with limitations in computational power to process large face databases and variations in face features even for a particular person, the desired efficiency and accuracy could not be achieved. As a result, the activities were dormant until the work in 1990 by Kirby and Sirovich [4] on low dimensional face representation which was derived using the Karhunen-Loeve transform. Further, the pioneering work of Turk and Pentland on eigenface [5] reinvigorated the face recognition research and the decades that followed saw the evolution of vast numbers of AFR algorithms.

The general term of face recognition can refer to different application scenarios. One scenario is called recognition or identification, and another is called authentication or verification. In either scenario, face images of known persons are initially enrolled into the system. This set of persons is sometimes referred to as the gallery. Later images of these or other persons are used as probes to match against images in the gallery. In a recognition scenario, the matching is one-to-many, in the sense that a probe is matched against all images of the gallery to find the best match above some threshold. In an authentication scenario, the matching is one-to-one, in the sense that the probe is matched against the gallery entry for a claimed identity, and the claimed identity is taken to be authenticated if the quality of match exceeds some threshold. The recognition scenario is more technically challenging than the authentication scenario. One reason is that in a recognition scenario a larger gallery tends to present more chances for incorrect recognition. Another reason is that the whole gallery must be searched in some manner on each recognition attempt. A third scenario may also arise where the test individual may or may not be in the system database. The query face image is compared against all the face images in the database, resulting in a score. A score higher than a given threshold may result in an alarm of recognition.

The techniques of AFR can be broadly divided into two interlinked operations: (a) face detection and (b) face recognition as shown in Figure 1.1. Face detection is a necessary initial step, with the purpose of localizing and extracting the face region from the background. The solution to the problem involves segmentation and extraction of faces and possibly facial features from an uncontrolled background. Face recognition operation involves performing verification and identification [6]. This stage takes the probe image extracted from the scene during the face detection stage and compares it with a database of previously enrolled known faces. Searching for the closest matching images is then carried out for identifying the most likely matched face. The final stage of face recognition is identification and verification. Identification is the process of comparing a face with a set of two or more faces in order to determine the most likely match. Face verification is the process of comparing the test face with another known face in the database, resulting in either acceptance as client or rejection of the face as imposture. The image which is to be either verified or recognized is said to be a query image. Gallery images are those images with which the query image is compared.

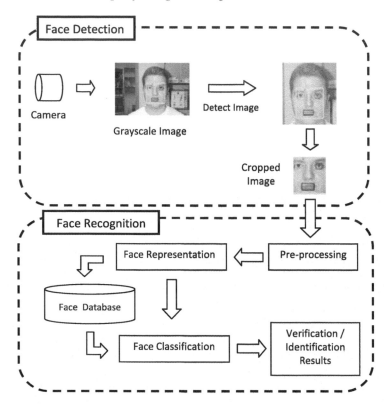

**FIGURE 1.1**: (a) System diagram of a typical face detection/face recognition system

Face recognition technology has significantly advanced since the time when the eigenface method was proposed. In the constrained situations, for example where lighting, pose, stand-off, facial wear and facial expression can be controlled, automated face recognition can provide high recognition rates, especially when the database (gallery) contains a large or even small number of face images. However, even in very controlled imaging conditions, such as those used for passport photographs, the reported error rate is high [7]. In less controlled environments, the performance degrades even further [8]. Training within a system under certain imaging conditions (single illumination, pose and motion pattern), and being able to recognize under arbitrary changes in these conditions, can be considered a challenging problem formulation. A common limitation of these methods is related to the requirement of fairly restrictive and labor-intensive training data acquisition protocol, in which a number of fixed views are collected for each subject and then appropriately labelled.

While performance of the systems commercially available is reasonable, it is questionable whether the face itself, without any contextual information, is a sufficient basis for recognizing a person from a large number of identities with an extremely high level of confidence. It is difficult to recognize a face from images captured from two drastically different views. Further, current face recognition systems impose a number of restrictions on how the facial images are obtained, sometimes requiring a simple background or special illumination. In order for the face recognition systems to be widely adopted, they should automatically detect whether a face is present in the acquired image; locate the face if there is one; and recognize the face from a general viewpoint. The challenge in vision-based face recognition is the presence of a high degree of variability in human face images. There can be potentially very large intra-subject variations (due to 3D head pose, lighting, facial expression, facial hair and ageing and rather small intersubject variations, due to the similarity of individual appearances.

Currently available vision-based recognition techniques can be mainly categorized into two groups, based on the face representations which are (i) appearance-based techniques which use holistic texture features, and (ii) geometry-based techniques which use geometrical features of the face. Experimental results have shown that the appearance-based methods generally perform better recognition tasks than those based on geometry, since it is difficult to robustly extract geometrical features in face images especially from low resolutions and poor quality images (i.e., to extract features under uncertainty). However, the appearance-based recognition techniques have their own limitations in recognizing human faces in images with wide variations of head poses and illumination.

In summary, external and internal facial components, distinctiveness, configuration and local texture of facial components all contribute to the process of face detection and recognition. In contrast, humans can seamlessly blend and independently perform appearance-based and geometry-based detection and recognition tasks most efficiently.

### 1.3.2   Process flow in face recognition system

The technique of automatic face recognition (AFR) can be explicitly elaborated with the help of three datasets. Dataset 1 contains the nonface image samples as shown in Figure 1.2 (a), dataset 2 contains face images as shown in Figure 1.2 (b), which are known to the system and dataset 3 contains face images, as shown in Figure 1.2 (c), unknown to the system. Basic steps of AFR can now be described in the following steps along with the pictorial interpretation:

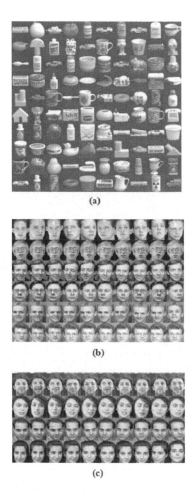

**FIGURE 1.2**: (a) Nonface datasets, (b) face dataset known to the system and (c) face dataset unknown to the system

- Mixed face and nonface images are input to the system as shown in Figure 1.3 and the detection module discriminates between the face images and nonface images.

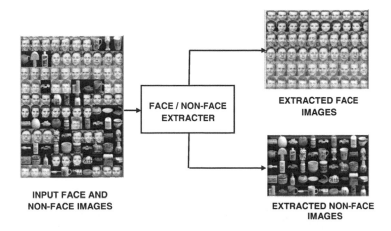

**FIGURE 1.3**: Face detection module

- The recognition phase may have two types of modules performing different discriminating functions: (1) The open test module, as shown in Figure 1.4, performs the process of verification and decides whether the face is known to the system or not; (2) The closed test module discriminates and decides the particular task of class labelling by a process of identification, as shown in Figure 1.5.

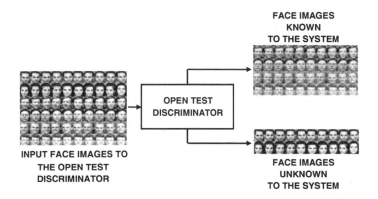

**FIGURE 1.4**: Open test discriminator module

A face recognition system generally consists of different functional modules performing functions of (a) face detection, (b) feature extraction and (c)

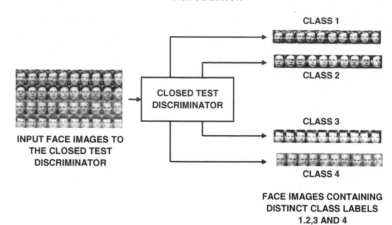

CLASS 1

CLASS 2

CLASS 3

CLASS 4

INPUT FACE IMAGES TO
THE CLOSED TEST
DISCRIMINATOR

CLOSED TEST
DISCRIMINATOR

FACE IMAGES CONTAINING
DISTINCT CLASS LABELS
1.2,3 AND 4

**FIGURE 1.5**: Closed test classification module

matching. The process of face detection basically deals with face localization and normalization. The functions performed under (a) and (b) may constitute the face detection module. Generally, matching and identification constitute the face recognition process. The function of localization may involve the segmentation of the face area from the background. This may also provide the location and scale of the face. Face features such as eyes, nose, mouth and facial outline are localized during this operation. However, the system is expected to recognize face images with varying pose and illumination. Therefore, some form of face image normalization is required to normalize the face image geometrically and photometrically. The geometrical normalization process transforms the face image into a standard frame. Warping or morphing may be used for more elaborate geometric normalization. The photometric normalization process normalizes the face image based on illumination level and gray scale. The next module performs the function of face feature extraction on the normalized face image and extracts salient information that is useful for distinguishing faces of different persons. The process needs to be robust with respect to geometric and photometric variations. The extracted face features are then used for matching by the face matching module, where the extracted features from the input face images are matched against one or more of the stored face images in the database. The main challenge in this stage of face recognition is to find a suitable similarity measure for comparing facial features. The accuracy of a face recognition system is highly dependent on the features that are extracted to represent a face, which, in turn, depend on the correct face localization and normalization.

### 1.3.3 Problems of face detection and recognition

Automated methods of face detection that use facial features as essential elements of discrepancy to determine identity have been studied for more than thirty years. Also the face recognition techniques, which involve classification of extremely confusing multi-dimensional input signals, and matching them with the known database, have been studied seriously. However, excessively large numbers of training face images are required for classifying an array representing a test face in high dimensions. Therefore the systems are prone to the so-called curse of dimensionality.

The human face is not a uniquely rigid object and there are numerous factors that cause the appearance of the face to vary. The sources of variation in the facial appearance can be categorized into two groups: intrinsic factors and extrinsic ones. Intrinsic factors are related to the physical nature of the face and are independent of the observer. Intrinsic factors can be further classified into two categories: intrapersonal and interpersonal [9]. Intrapersonal factors are responsible for varying the facial appearance of the same person due to age, facial expression and facial paraphernalia such as facial hair, glasses, cosmetics, etc. Interpersonal factors, however, are responsible for the differences in the facial appearance of different people, some examples being ethnicity and gender. Extrinsic factors cause the appearance of the face to alter via the interaction of light with the face and the observer. These factors include illumination, pose, scale and imaging parameters such as resolution, focus, noise, etc. These factors are reflected negatively in the techniques and technologies of face detection and recognition systems having acceptable performance in a real-time scenario. Mainly five key factors may be identified which have to be addressed in any viable technique. They are:

1. Several 2D methods do well in performing recognition tasks only under moderate illumination variations within a given range, while performances noticeably degrade at large variations in illumination conditions.

2. Occlusions can dramatically affect face recognition performances, in particular if they are located on the upper-side of the face.

3. Pose changes such as head rotation also affect the identification process, because they introduce projective deformations and self-occlusion. This problem is accentuated where security cameras change the viewing angles when they are outside of the range of the designed viewing angle of a system.

4. Sometimes, even under an acceptable viewing angle, extreme expression changes of the face may result in false recognition.

5. Another important factor is related to the change of face over a period of time, as the shape of face changes in a nonlinear way due to ageing. This

problem is harder to solve and not much work has been done especially in accounting for age variations.

### 1.3.4 Liveness detection for face recognition

Liveness detection, which aims at recognition of human physiological activities as the liveness indicator to prevent spoofing attacks, is becoming a very active topic. The most common faking way is to use a facial photograph of a valid user to spoof face recognition systems. Nowadays, video of a valid user can also be easily captured by a needle camera for spoofing. Therefore, anti-spoof solutions also form a part of face recognition systems.

In general, there are three ways to spoof a face recognition system by using a photograph of a valid user or by using a video of a valid user or by using a 3D model of a valid user. Photo attack is the cheapest and easiest spoofing approach, since one's facial image is usually very easily available in the public domain, for example, can be downloaded from the web, or captured unknowingly by a camera. The impostor can rotate, shift and bend the photo before the camera like a live person to fool the authentication system. It is still a challenging task to detect whether an input face image is from a live person or from a photograph.

In general, a human is able to distinguish a live face and a photograph without any effort, since a human can very easily recognize many physiological clues of liveness, for example, facial expression variation, mouth movement, head rotation, eye change. However, the tasks of computing these clues are often complicated for the computer, even impossible for some clues under the unconstrained environment.

From the static view, an essential difference between a live face and a photograph is that a live face is a fully three- dimensional object while a photograph could be considered as a two dimensional planar structure. With this natural trait, the structural changes due to motion is employed [10] to yield the depth information of the face for the detection of a live person or a still photo. The disadvantages of depth information are that, first, it is hard to estimate depth information when the head is still. Second, the estimate is very sensitive to noise and lighting condition and therefore is not reliable.

Compared to photographs, another prominent characteristic of live faces is the occurrence of the non-rigid deformation and appearance change, such as mouth motion and expression variation. The accurate and reliable detection of these changes usually needs high quality input data or user collaboration. The optical flow technique which uses input video to obtain information of face motion for liveness judgement [11] is used, but it is vulnerable to photo motion in depth and photo bending. Some researchers use the multi-modal approaches of face-voice against spoofing [12], [13], exploiting the lip movement during speaking. This kind of method needs a voice recorder and user collaboration. An interactive approach is tried [12], requiring the user to react to an obvious response of head movement.

Additionally, Fourier spectra is applied to classify live faces or faked images, based on the assumption that the high frequency components of the photo are less than those in live face images [14]. But using thermal infrared imaging cameras, face thermogram also could be applied for liveness detection [15].

## 1.4 Tests and metrics

The standard protocol in evaluating face recognition algorithms requires three separate sets of images: training, gallery and probe sets. The training set is for learning whereas the gallery and probe sets are used for testing the recognition algorithm. The gallery set contains images with known identities while the identities in the probe set are unknown. As the complexity of the identification depends on the number of individuals in the gallery and probe sets, this number should be large. In principle, there should be no overlap between training and testing images, not only in terms of identity but also in terms of other physical conditions. Ideally, to ensure that the system is not tuned to any specific condition, training and testing image sets should originate from different and independent sources. Unfortunately, due to various conditions these practices are difficult to follow.

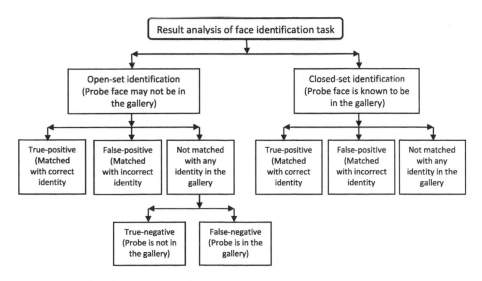

**FIGURE 1.6**: Possible results of identification operation

After detection, all face recognition systems are basically required to perform the tasks of identification and verification. During these processes, several results are possible as shown in Figure 1.6 and Figure 1.7.

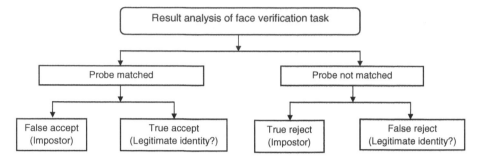

**FIGURE 1.7**: Possible results of verification operation

These possible results are related to important metrics in recognizing a probe face image as given under:

1. False acceptance rate (FAR) defines a metric when an impostor is accepted as the gallery face and this condition occurs when the similarity of the impostor template falls within the intrauser variation of a genuine gallery face;

2. False rejection rate (FRR) is a metric when a gallery subject is rejected as impostor and this condition may occur when the gallery face is of poor quality;

3. False identification rate (FIR) is a metric which occurs due to mis-recognition between two gallery faces.

It may be noted that these three kinds of errors are quite different and simply taking error rate as the measure of the performance may not be a good choice. Without loss of generality, it may be assumed that false rejection is more serious than false identification, and the false acceptance is the most serious error.

However, in a face recognition system, FRR and FAR are related to reference threshold, which is defined as a value that can decide whether a person is genuine or impostor. In other words, basically the value of reference threshold authenticates a person as genuine or an impostor. Depending upon the application, the system can be tuned for desired value of FAR and FRR, by selecting the threshold value. Figure 1.8 explains the effect of tuning of threshold value $R_{th}$ on false acceptance and false rejection, where the genuine and impostor distributions are represented by typical probability distribution curves. Sliding of $R_{th}$ value to $R_{th1}$ leads to the decrease in false rejection but increase in false acceptance. Similarly, sliding of $R_{th}$ to $R_{th2}$ leads to the decrease in false acceptance but increase in false rejection. Therefore, tuning of reference threshold is very important, otherwise, it may lead to false acceptance or false rejection.

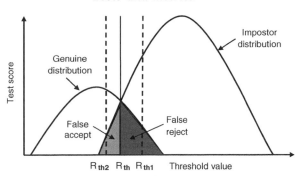

**FIGURE 1.8**: Effect of change in threshold value on FRR and FAR

The performance of a face recognition system may also be summarized using other single valued measures called equal error rate (ERR) and d-prime value. The EER is the point where the false reject rate equals the false accept rate, and is the most commonly stated single number from the receiver operating characteristics (ROC) curve. A lower ERR value indicates better performance of the system. The detection error trade-off curve plots the FRR against the FAR at various thresholds on a normal deviate scale and interpolates between these points. When a linear, logarithmic or semilogarithmic scale is used to plot the error rates, then the resulting graph is the ROC. The ROC curve summarizes the percent of a set of probes that is falsely rejected as a trade-off against the percent that is falsely accepted. The ROC-based method, however, is focused on binary classification problems, while face recognition is inherently a multiclass problem. Extending ROC methods to multiclass is non-trivial.

Sometimes, a measure is used which combines FAR and FRR into the decision cost function (DCF) and is given by

$$DCF = C_{FR}P_{gal}FRR + C_{FA}P_{imp}FAR \qquad (1.1)$$

where $C_{FR}$ is the cost of false rejection, $C_{FA}$ is the cost of false acceptance, $P_{gal}$ is the a priori probability of gallery face image and $P_{imp}$ is the a priori probability of impostors. In a particular case, when the probabilities are 0.5 and the cost is 1, DCF is termed as half total error rate (HTER) and is given by,

$$HTER = (FRR + FAR)/2 \qquad (1.2)$$

Another measure of performance is the d-prime value. This value measures the separation between the means of the probability distributions in standard deviation units of genuine gallery faces and impostor faces. A further metric used for face recognition systems, which is running over a period of time, is called the ability to verify rate (ATV). This rate is a combination of failure to enrol and false mismatch rates and indicates the overall percentage of users who are capable of authenticating on day to day basis. This metric represents the group of users who cannot enrol along with users falsely rejected by the system. A high ATV is desirable, though no system can have 100 percent ATV.

---

## 1.5  Cognitive psychology in face recognition

Many investigations by the cognitive psychology and neurophysiology field related to face recognition address the basic question of how do people recognise faces. More importantly, what are the processes for recognizing faces? The main issue is concerned with the empirical evaluation and theoretical development of cognitive processes that enables conceptual and practical understanding of how the human recognizes faces. Such investigations would lead, expectedly, better machine translation of the face recognition process without misclassification and ambiguity under various constraints. In terms of physiology such studies may be able to investigate into prosopagnosia, where the patients are no longer able to recognise faces of previously known individuals and the effects of face distinctiveness, when recognising faces. Other studies have also been carried out in the use of face caricatures, which distort faces to improve their uniqueness and distinction amongst the general population.

The human visual system is believed to have rudimentary preference for face-like patterns. While studying the development of face recognition capability from early childhood, it has been observed that the preference for faces and face-like patterns for a child occurs hours after birth and starts with the mother's face. According to the predictions of the intersensory redundancy hypothesis (IRH), during early development, perception of faces is enhanced in unimodal visual stimulation (i.e., silent dynamic face) rather than bimodal audiovisual mode(i.e., dynamic face with synchronous speech) stimulation [16],[17]. It is still unclear how face perception and recognition capability is developed from the preferences for face and face-like patterns into the abilities of face recognition by adults by correlating the face recognition expertise gained over the years.

In later years the human expertise in recognizing a face is related to the ability to process and recognize faces using efficient processing styles, such as featural, configural and holistic. Featural processing refers to the perception and recognition of faces on the basis of the individual features themselves, such as the shape of the eyes or the size of the nose. Configural processing [18], [19], a more advanced processing style than featural, refers to the perception and recognition of faces on the basis of not only the features but also the spacing between features. Holistic processing refers to the perception and recognition of faces as a whole rather than based on parts of the face. It is generally believed that infants process faces using immature styles, such as featural processing [20]. On the other hand, teens have more experience with faces and are thought to process them on the basis of configural information rather than featural information alone [21]. Adults, through their expertise, experience perceiving and recognizing faces as well as advanced processing of faces, though adults show a remarkable deficit in recognition of inverted faces. Gradually, a more sophisticated holistic strategy involving configural information helps in achieving ability for robust face recognition. This pattern of behavior suggests that over the course of several years, a shift in strategy occurs.

Another issue experienced while developing the capability of face recognition is known as other-race effect. The other-race effect refers to individuals being better able to discriminate faces within their own race than within another race. As one gains experience with faces within their own race, discrimination of faces within other races diminishes. A perceptual narrowing effect also occurs for the perception of faces [22], when face perception narrows as a result of increased experience with faces. This effect eventually leads to development of a so-called face prototype. A face prototype refers to an average of numerous faces which is perceived by adults and supposed to have more discriminative power than any one of the individual faces.

From a neurophysiological point of view, it seems that the human visual system appears to devote specialized neural resources for face perception and recognition. It has been suspected that unique cognitive and neural mechanisms may exist for face processing in the human visual system. Indeed, there is a great deal of evidence that the primary locus for human face processing may be found in the extrastriate visual cortex. This region shows an intriguing pattern of selectivity and generality which is evident from the fact that schematic faces do not give rise to much activity and animal faces do elicit a good response. The idea of the existence of a dedicated face processing module appears very strong.

It is further suggested that facial identity and expression might be processed by separate neurophysiological systems. If this corollary is established then it is possible to extract a facial expression independently of the identity and vice versa. The computational implications of this question would determine whether a biologically based implementation would be able to identify a person without taking into account the person's expression. In such a scenario, it will be possible to judge the facial emotions in a humancomputer interaction application without going through the process of extracting a representation of identity. However, at the same time, it must be pointed out that although there seems to be a significant amount of dissociation between identity and expression, most studies do leave some room for overlap, perhaps at the representational stage. For example, although some neurons responded only to identity and some only to expression, a smaller subset of neurons responded to both factors.

However, while developing an automated face recognised system, researchers often investigate face expertise by assessing perception of first- and second- order relations. First-order relations refer to the basic features of a face and second-order relations refer to the spacing between features within a face. Perception or recognition of a face on the basis of first-order relations is prone to errors in recognition, while the perception or recognition of a face on the basis of second-order relations is more robust [23].

# Chapter 2

# Face detection and recognition techniques

## 2.1 Introduction to face detection

Face detection is a necessary first step in face recognition systems, with the purpose of localizing and extracting the face region from the background. It also has several applications in areas such as content-based image retrieval, video coding, video conferencing, crowd surveillance and intelligent human computer interfaces.

The human face is a dynamic object and has a degree of variability in its appearance. This makes face detection a difficult problem, particularly when integrated with computer vision systems. A wide variety of techniques have been proposed, ranging from simple edge-based algorithms to composite high-level approaches utilizing advanced pattern recognition methods. With over 150 reported approaches to face detection, the research in face detection (Figure 2.1) gives different approaches associated with face detection technology. Many of the current face recognition techniques assume the

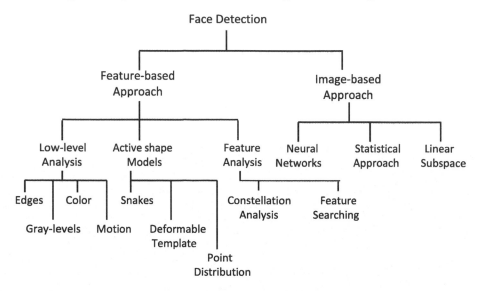

**FIGURE 2.1**: Different approaches of face detection techniques

**FIGURE 2.2**: Typical images used for face classification

availability of frontal faces of similar sizes. In reality, this assumption may not hold due to the varied nature of face appearance and environmental conditions. Figure 2.2 shows some typical test images used for face detection. In realistic application scenarios such as the example in Figure 2.3, a face could occur in a complex background and in many different positions. Detection and recognition systems that are based on face images such as in Figure 2.2, are likely to make false detection and recognition, as some areas of the background may be mistaken as a face. In order to rectify the problem, a visual front-end processor is needed to localize and extract the face region from the background.

**FIGURE 2.3**: Realistic face detection problem

Given an arbitrary image, the goal of the face detection process is to determine whether or not there is any face in the image and, if present, return the image location and extent of each face. The challenges associated with the face detection process can be attributed to the following factors:

- **Occlusion**. Faces may be partially occluded by other objects. In an image with a group of people, some faces may partially occlude other faces.

- **Facial expression**. The appearances of faces are directly affected by a person's facial expression.

- **Pose**. The images of a face vary due to the relative camera-face pose such as frontal, 45 degree, profile, upside down etc.

- **Illumination**. When a face image is formed, factors such as lighting conditions and camera characteristics also affect the appearance of a face.

## 2.2 Feature-based approaches for face detection

Numerous methods have been proposed to first detect facial features and then to infer the presence of a face. Facial features such as eyebrows, eyes, nose, mouth and hairline are commonly extracted. Based on the extracted features, a statistical model is built to describe their relationships and to verify the existence of a face. One problem with these feature-based algorithms is that the image features can be severely corrupted due to illumination, noise and occlusion.

The systems based on the feature-based approach can be further divided into three groups. (a) Given a typical face detection problem in locating a face in a cluttered scene, low-level analysis is executed which first deals with the segmentation of visual features using pixel properties such as gray scale and colour. Because of the low-level nature, features generated from this analysis are ambiguous. (b) In feature analysis systems, visual features are organized into a more global concept of the face using face geometry information. Through feature analysis, feature ambiguities are reduced and locations of the face and facial features are determined. (c) The last group involves the use of active shape models. These models ranging from snakes, proposed in the late 1980s, to the more recent point-distributed models (PDM) have been utilized for the purpose of complex and non-rigid feature extraction and tracking such as eye pupil and lip tracking.

## 2.2.1 Low-level analysis

### 2.2.1.1 Edges

Edge representation for feature extraction was applied in the earliest face detection work [24]. The work was based on analyzing line drawings of the faces from photographs and aims to locate facial features. This work was later modified [25] by proposing a hierarchical framework to trace an outline of a human head. The work includes a line-follower implemented with the curvature constraint to prevent it from being distracted by noisy edges. Edge features within the head outline are then subjected to feature analysis using shape and position information of the face.

More recent examples of edge-based techniques can be found in [26] for facial feature extraction and in [27] for face detection. In an edge-based approach to face detection, edges need to be labeled and matched to a face model in order to verify correct detections. Some times this is accomplished by labeling edges as the left side, hairline or right side of a front view face and is followed by matching these edges against a face model, using the golden ratio for an ideal face. Further, edge-based techniques have also been applied for detecting glasses in facial images [28]. Many different types of edge operators have been applied. The Sobel operator, [26] and MarrHildreth edge operator [29] are some of the operators used as edge detectors. A variety of first and second derivatives (Laplacian) of Gaussians have also been used in the other methods. For instance, a Laplacian of large scale was used to obtain a line drawing in [24] and steerable and multiscale orientation filters [30]. The steerable filtering consists of three sequential edge detection steps which include detection of edges, determination of the filter orientation of any detected edges and stepwise tracking of neighboring edges using the orientation information. The algorithm has allowed an accurate extraction of several key points of the eye.

### 2.2.1.2 Gray-level analysis

Besides edge details, the gray information within a face can also be used as features. Facial features such as eyebrows, pupils and lips appear generally darker than their surrounding facial regions. This property is exploited to differentiate various facial parts. Several recent facial feature extraction algorithms [31] search for local gray minima within segmented facial regions. In these algorithms, the input images are first enhanced by contrast-stretching and gray-scale morphological routines to improve the quality of local dark patches and thereby make detection easier. The extraction of dark patches is achieved by low-level gray-scale thresholding. Some algorithms make use of a weighted human eye template to determine possible locations of an eye pair. Local maxima, which are defined by a bright pixel surrounded by eight dark neighbours, are used to indicate the bright facial spots such as nose tips [32]. The detection points are then aligned with feature templates for correlation measurements.

Yang and Huang [33], on the other hand, explore the gray-scale behaviour of faces in mosaic (pyramid) images. When the resolution of a face image is reduced gradually either by subsampling or averaging, macroscopic features of the face disappear and at low resolution, face region becomes uniform. Starting at low resolution images, face candidates are established by a set of rules that search for uniform regions. The face candidates are then verified by the existence of prominent facial features using local minima at higher resolutions. The technique is incorporated into a system for rotation invariant face detection [34] and an extension of the algorithm is presented in [35].

### 2.2.1.3 Color information in face detection

Whilst gray information provides the basic representation for image features, colour is a more powerful tool for discerning object appearance. Due to the extra dimensions that colour has, two shades of similar gray information might appear very differently in a colour space. It is also found that different human skin colour gives rise to a tight cluster in colour spaces, even when faces of difference races are considered. This means that colour composition of human skin differs little across individuals.

One of the most widely used colour models is RGB representation in which different colours are defined by combinations of red, green and blue primary colour components. Since the main variation in skin appearance is largely due to luminance change (brightness), normalized RGB colours are generally preferred, so that the effect of luminance can be filtered out. Besides RGB colour models, there are several other alternative models currently being used in face detection research. In [36] HSI colour representation has been shown to have advantages over other models in giving large variance among facial feature colour clusters. Hence this model is used to extract facial features such as lips, eyes and eyebrows. Since the representation strongly relates to human perception of colour, it is also widely used in face segmentation schemes.

The YIQ colour model has also been applied to face detection [37]. By converting RGB colours into YIQ representation, it was found that the I-component, which includes colours ranging from orange to cyan, enhances the skin region of Asians. The conversion also effectively suppresses the background of other colours and allows the detection of small faces in a natural environment. Other colour models applied to face detection include HSV, YES, YCrCb, YU, CIE-xyz, CSN (a modified RQ representation) and UCS. Terrilon et al. [38] recently presented a comparative study of several widely used colour spaces (or more appropriately named chrominance spaces) for face detection. A general conclusion is that the most important criterion for face detection is the degree of overlap between skin and non-skin distributions in a given space (and this is highly dependent on the number of skin and nonskin samples available for training).

Colour segmentation can basically be performed using appropriate skin colour thresholds where skin colour is modeled through histograms or charts. More complex methods make use of statistical measures that model face variation within a wide user spectrum. For instance, a Gaussian distribution characterized by its mean and covariance matrix, is employed to represent a skin colour cluster of thousands of skin colour samples taken from difference races [39],[40]. Incidentally, colour detection can be more robust against changes in environment factors such as illumination conditions and camera characteristics.

### 2.2.1.4   Motion-based analysis

Motion information is a convenient means of locating a moving face when video is available. A straightforward way to achieve motion segmentation is by frame difference analysis. This approach, whilst simple, is able to discern a moving foreground efficiently regardless of the background content. In [41], moving silhouettes that include face and body parts are extracted by thresholding of accumulated frame difference. In [25], the existence of an eye-pair is hypothesized by measuring the horizontal and the vertical displacements between two adjacent candidate regions obtained from frame difference. Another way of measuring visual motion is through the estimation of moving contours of the face. Compared to frame difference, results generated from moving contours are always more reliable, especially when the motion is insignificant. Unlike the methods described above which identify moving edges and regions, these methods rely on the accurate estimation of the apparent brightness velocities called optical flow. Because the estimation is based on short-range moving patterns, it is sensitive to fine motion.

## 2.2.2 Active shape model

Unlike the face models described in the previous sections, active shape models depict the actual physical and, hence, higher-level appearance of features. Once released within a close proximity of a feature, an active shape model interacts with local image features (edges, brightness) and gradually deforms to take the shape of the feature. There are generally three types of active shape models in the contemporary facial feature extraction. The first type uses a generic active contour (snakes) [42]. To take into account the a priori of facial features and for a better performance of snakes, deformable templates are introduced. A new generic flexible model, which is termed smart snakes, is also introduced to provide an efficient interpretation of the human face [43]. The model is based on a set of labelled points that are only allowed to vary to certain shapes according to a training procedure.

Active contours are commonly used to locate a head boundary. The snake is first initialized at the proximity around a head boundary which locks onto nearby edges and subsequently assume the shape of the head. The typical natural evolution in snakes is shrinking or expanding and by that process the contours deviate from the natural evolution and eventually assume the shape of a head boundary at a state of equilibrium. Two main considerations in implementing a snake are the selection of the appropriate energy terms and the energy minimization technique. Elastic energy is used commonly as internal energy. It is proportional to the distance between the control points on the snake and therefore gives the contour an elastic-band characteristic that causes it to shrink or expand. The external energy consideration is dependent on the type of image features considered. In addition to gradient information, the external energy term includes a skin colour function which attracts the contour to the face region. Energy minimization can be achieved by optimization techniques such as the steepest gradient descent.

Even though snakes are generally able to locate feature boundaries, their implementation is still plagued by two problems. Part of the contour often becomes trapped onto false image features. Furthermore, snakes are not efficient in extracting nonconvex features due to their tendency to attain minimum curvature. These problems are addressed by introducing a parameterized snake model for face and head boundary extraction. The parameterized model biases the contours toward the target shape and thereby allows it to distinguish false image features and not be trapped by them. Once the contours reach equilibrium, the model is removed and the contours are allowed to act individually as a pair of conventional snakes, which leads to the final boundary extraction.

Applying snake models in locating a facial feature boundary is not an easy task because the local evidence of facial edges is difficult to organize into a sensible global entity using generic contours. The low brightness contrast around some of these features also makes the edge detection process problematic. The concept of snakes is used by incorporating global information of the eye to improve the reliability of the extraction process [44].

A deformable eye template based on salient features of eye is parameterized snake. The evolution of a deformable template is sensitive to its initial position because of the fixed matching strategy. The processing time is also very high due to the sequential implementation of the minimization process. The weights of the energy terms are heuristic and difficult to generalize [45]. Improvement of the eye template and mouth template matching leads to high accuracy by trading off the considerations of some parameters that have lesser effects on the overall template shape. In a more recent development [46], eye corner information is used to estimate the initial parameters of the eye template. The inclusion of this additional information has allowed more reliable template fitting.

### 2.2.3   Feature analysis

Features generated from low-level analysis are likely to be ambiguous. For instance, in locating facial regions using a skin colour model, background objects of similar colour can also be detected. This is a many-to-one mapping problem which can be solved by higher level feature analysis. The knowledge of face geometry has been employed to characterize and subsequently verify various features from their ambiguous state. There are two approaches in the application of face geometry. The first approach involves sequential feature searching strategies based on the relative positioning of individual facial features. The confidence of a feature existence is enhanced by the detection of nearby features. The techniques in the second approach group features as flexible constellations using various face models.

Feature searching techniques begin with the determination of prominent facial features. Then other less prominent features are hypothesized using anthropometric measurements of face geometry. For instance, a small area on top of a larger area in a head and shoulder sequence implies a face on top of shoulder scenario, and a pair of dark regions found in the face area increases the confidence of a face existence. The pair of eyes is the most commonly applied reference feature for its distinct side-by-side appearance. Other features, including a main face axis, outline (top of the head) and body (below the head) have also been used [47].

The facial feature extraction algorithm starts by hypothesizing the top of a head and then a searching algorithm scans downward to find an eye-plane which appears to have a sudden increase in edge densities, measured by the ratio of black to white along the horizontal planes. The length between the top and the eye-plane is then used as a reference length. Using this reference length, a flexible facial template covering features such as the eyes and the mouth is initialized on the input image. The flexible template is then adjusted to the final feature positions using an edge-based cost function. A system for face and facial feature detection which is also based on anthropometric measures is also proposed [48]. In this system, possible locations of the eyes are established in binarized pre-processed images. Then the algorithm goes on to search for a nose, a mouth and eyebrows. Each facial feature has an associated evaluation function, which is used to determine a most likely face candidate, weighted by their facial importance with manually selected coefficients.

An automatic facial features searching algorithm called GAZE is proposed [30] based on the motivation of eye movement strategies. The heart of the algorithm is a local attentive mechanism that foveated sequentially on the most prominent feature location. A multilevel saliency representation is first derived using a multi-orientation Gaussian filter. The most prominent feature is extracted using coarse to fine evaluation on the saliency map. Other facial regions like the nose and the mouth are also detected at the later iterations. Because the test images used by the algorithm contain faces of different orientation (some faces are tilted) and slight variation in illumination conditions and scale, the high detection rate indicates that this algorithm is relatively independent of those image variations.

Eye movement strategies are also considered as a basis of several feature analysis algorithms. In an algorithm, description of the search targets (the eyes) is constructed by averaging Gabor responses from a retinal sampling grid centered on the eyes of the subjects in the training set [49]. The smallest Gabor functions are used at the center of the sampling grid, while the largest are used off-center. For detecting the eyes, a saccadic search algorithm is applied which consists of initially placing the sampling grid at a random position in the image and then moving it to the position where the Euclidian distance between the node of the sampling grid and the node in the search target is the smallest. The grid is moved around until the saccades become smaller than a threshold.

## 2.2.4   Image-based approaches for face detection

Although some of the recent feature-based attempts have improved the ability to cope with the unpredictability, most are still limited to the detection of head and shoulder and quasi-frontal faces. There is still a need for techniques that can perform in more hostile scenarios such as detecting multiple faces with clutter-intensive backgrounds. This requirement has prompted the use of image-based approaches for face detection. In most of the image-based approaches, specific application of face knowledge is avoided. This eliminates the potential of modelling error due to incomplete or inaccurate face knowledge. The basic approach in detection of the face is via a training procedure which classifies examples into face and non-face prototype classes. Comparison between these classes and a 2D intensity array extracted from an input image allows the decision of face existence to be made. The simplest image-based approaches rely on template matching [50], but these approaches do not perform well and more complex techniques are proposed.

Most of the image-based approaches apply a window scanning technique for detecting faces, where exhaustive search of the input image for possible face locations is carried out on all scales. However, there are variations in the implementation of this algorithm for almost all the image-based systems. Typically, the size of the scanning window, the sub-sampling rate, the step size and the number of iterations vary depending on the method proposed and the need for a computationally efficient system.

### 2.2.5 Statistical approaches

Based on the work of maximum likelihood face detection, a statistical method is proposed where Kullback relative information (Kullback divergence) is used [51]. The training procedure results in a set of look-up tables with likelihood ratios. To further improve performance and reduce computational requirements, pairs of pixels which contribute poorly to the overall divergency are dropped from the look-up tables and not used in the face detection system. By including error bootstrapping, the technique was incorporated in a real-time face tracking system [52].

Other methods of face detection, such as linear subspace methods and neural networks, can also be used for face recognition. These techniques are discussed under the section on face recognition.

---

## 2.3 Face recognition methods

During the last 20 years, numerous face recognition algorithms have been proposed. It appears that, among several techniques, the subspace methods or appearance-based methods contributed significantly to the development of face detection and recognition techniques. Based on the type of image transformation used, the subspace methods may again broadly be divided into linear and non linear methods.

All existing face recognition techniques can be classified into five types based on the way they identify the face. They are

1. Appearance-(feature) based which uses all-inclusive texture features.

2. Subspace-based face recognition

3. Techniques using neural networks.

4. Model-based which works shape and texture of the face, along with 3D depth information.

5. Other techniques such as correlation techniques and the use of support vector machines.

## 2.3.1    Geometric feature-based method

The geometric feature-based approaches are the earliest approaches to face recognition and detection. In these systems, the significant facial features are detected and the distances among them as well as other geometric characteristics are combined in a feature vector that is used to represent the face. To recognize a face, first the feature vector of the test image and of the image in the database is obtained. Second, a similarity measure between these vectors, most often a minimum distance criterion, is used to determine the identity of the face. The template based approaches will outperform the early geometric feature based approaches. The template-based approaches represent the most popular technique used to recognize and detect faces. Unlike the geometric feature-based approaches, the template-based approaches use a feature vector that represent the entire face template rather than the most significant facial features.

## 2.3.2    Subspace-based face recognition

In general, subspace methods use a training set of face images in order to compute a coordinate space in which face images are compressed to fewer dimensions, whilst maintaining maximum variance across each orthogonal subspace direction. Images of faces, being similar in overall configurations, are not randomly distributed in this huge image space and thus can be described by a relatively low dimensional subspace. The main idea is to find the vectors that best account for the distribution of face images within the entire image space. These vectors of reduced dimension define the subspace of face images, which are referred to as *face space* [53]. Thus after the linearization the mean vector is calculated, among all images, and subtracted from all the vectors, corresponding to the original faces. The covariance matrix is then computed, in order to extract a limited number of its eigenvectors, corresponding to the greatest eigenvalues. These few eigenvectors, also referred to as *eigenfaces*, represent a base in a low-dimensionality space. When a new image has to be tested, the corresponding eigenface expansion is computed and compared against the entire database, according to such a distance measure (usually the Euclidean distance).

Many reviews on classical subspace based techniques have appeared during the last few years (see Table 2.1). Their study also reflects numerous important observations, prospects and constraints of these methods on various face datasets. Robust face detection and recognition schemes require low-dimensional feature representation for data compression purposes and also demand enhanced discrimination abilities for subsequent image classification. The representation methods usually start with a dimensionality reduction procedure, since the high dimensionality of the original space makes the statistical estimation very difficult, if not impossible. Moreover, the high-dimensional space is mostly empty.

**TABLE** 2.1: Highlights of some review articles on subspace-based face recognition techniques

| Researchers | Review articles related to |
|---|---|
| Ming-Hsuan Yang [54] | Comparative studies between Kernel Eigenfaces and Kernel Fisherfaces |
| Phillips et al. [7] | Review on subspace methods on FERET dataset |
| Zhao et al. [55] | Survey on still and video imagery |
| Baek et al. [56] | Comparative studies on PCA and ICA |
| Liu et al. [57] | Studies on kernel Fisher Discriminant methods |
| Roweis et al. [58] | Nonlinear dimensionality analysis |
| Shakhnarovich et al. [59] | Review on subspace method |
| Lu [60] | Review on ICA, PCA, KPCA Methods |
| Martinez et al. [61] | Comparative studies between PCA and LDA |
| Navarrete et al. [62] | Comparative analysis of subspace methods |
| Wang et al. [63] | Unified framework design |
| Delac et al. [64] | Comparative study on different methods which use DIFS and/or DFFS |

The subspace methods typically use some form of dimensionality reduction method such as principal component analysis (PCA), also referred to as the discrete Karhunen Loeve expansion. The Fisherface (FLD) method uses linear discriminant analysis (LDA) to produce a subspace projection matrix or the independent component analysis (ICA) method. The linear discriminant analysis has been proposed as a better alternative to the PCA. It expressly provides discrimination among the classes, while the PCA deals with the input data in their entirety, without paying any attention to the underlying structure. Indeed the main aim of the LDA consists in finding a base of vectors providing the best discrimination among the classes, trying to maximize the between-class differences and minimizing the within-class ones. These techniques are powerful tools in the field of face detection and recognition. In some approaches, such as the Fisherfaces, the PCA is considered as a preliminary step in order to reduce the dimensionality of the input space, and then the LDA is applied to the resulting space, in order to perform the real classification. In some cases the LDA is applied directly on the input space. Alternatively, a hybrid between the direct LDA and the fractional LDA, a variant of the LDA, can be tried in which weighed functions are used to avoid misclassification of face images.

Kernel PCA (KPCA) is the reformulation of traditional linear PCA in a high-dimensional space that is constructed using a kernel function. Kernel PCA computes the principal eigenvectors of the kernel matrix, rather than those of the covariance matrix, and the kernel space offers the nonlinear mappings. Some other contributions on non linear subspace methods are stated in Table 2.3.

The main disadvantage of the PCA, LDA and Fisherfaces is their linearity. Particularly the PCA extracts a low-dimensional representation of the input data only exploiting the covariance matrix, so that no more than first- and second-order statistics are used. It has been shown that first- and second-order statistics hold information only about the amplitude spectrum of an image, discarding the phase-spectrum, while some experiments bring out that the human capability in recognizing objects is mainly driven by the phase-spectrum. This is the main reason for which the independent component analysis (ICA)is introduced as a more powerful classification tool for the face recognition problem. The ICA can be considered as a generalization of the PCA, but providing three main advantages: (1) it allows a better characterization of data in an n-dimensional space; (2) the vectors found by the ICA are not necessarily orthogonals, so that they also reduce the reconstruction error; (3) they capture discriminant features not only exploiting the covariance matrix, but also considering the high-order statistics.

Highlights of some important reviews and surveys on subspace based face recognition are indicated in Table 2.1. Major contributions on linear and nonlinear subspace techniques are also shown in Tables 2.2 and 2.3.

**TABLE 2.2**: Major contributions on linear subspace methods in face recognition

| Subspace methods | Researchers |
|---|---|
| Principal component analysis (PCA) | Kirby et al. [4],1990 |
| PCA, Eigenface recognition | Turk et al. [5], Moghaddam et al. [65], 1991, 2004 |
| Fisher LDA PCA-LDA | Belhumeur et al. [66],1997 |
| Independent component analysis (ICA) | Barlett et al. Draper et al.[67], Wechesler et al. [68], 1996-2004 |
| Dimensional PCA (2DPCA) | Zhang et al. [33], 2003-2004 |
| Tensor faces | Niyogi et al. [69], 2005 |

### 2.3.3   Neural network-based face recognition

A further nonlinear solution to the face recognition problem is given by the neural networks. The advantage of the neural networks in classification over linear ones is that they can reduce misclassifications among the neighborhood classes. The basic idea is to consider a net with a neuron for every pixel in the image. Nevertheless, because of the pattern dimensions neural networks are not directly trained with the input images, but they are preceded by the

**TABLE 2.3**: Major contribution in nonlinear subspace methods in face recognition

| Subspace Methods | Researchers | Year |
|---|---|---|
| Kernel PCA(KPCA) | K. Muller, kim et al. [70] | 1998 |
| kernel Fisher discriminant (KFD) | Ming-Hsuan, Yang [54] | 2002 |
| Laplacian Faces | Hong-Jiang Zhang et al. [71] | 2005 |
| Local linear embedding (LLE) | Roweis et al. [58] | 2000 |
| Isomap | Ming-Hsuan, Yang [72] | 2002 |

application of such a dimensionality reduction technique. A solution to this problem is the use of a second neural network that operates in auto-association mode. At first, the face image, is approximated by a vector with smaller dimensions by the first network (auto-association), and then this vector is finally used as input for the classification net.

In general, the structure of the network is strongly dependent on its application field, so that different contexts result in quite different networks. A class of neural network known as the self-organizing map (SOM) is also tested in face recognition, in order to exploit their particular properties. SOM is invariant with respect to minor changes in the image sample, while convolutional networks provide a partial invariance with respect to rotations, translations and scaling. Probabilistic decision-based neural network is used in face detection, in eye localizers and also in face recognition. The flexibility of these networks is due to their hierarchical structure with nonlinear basis functions and a competitive credit assignment scheme. A hybrid approach is also worth mentioning, in which, through the PCA, the most discriminating features are extracted and used as the input of a radial basis function (RBF) neural network. The RBFs perform well for face recognition problems, as they have a compact topology and learning speed is fast.

However, in general, neural network-based approaches encounter problems when the number of classes increases. Moreover, they are not suitable for a single model image recognition task, because multiple model images per person are necessary in order to train the system.

## 2.3.4    Correlation-based method

Correlation based methods for face detection are based on the computation of the normalized cross correlation coefficient. The first step in these methods is to determine the location of the significant facial features such as eyes, nose or mouth. The importance of robust facial feature detection for both detection and recognition has resulted in the development of a variety of different facial feature detection algorithms. The facial feature detection method uses a set of templates to detect the position of the eyes in an image, by looking for the maximum absolute values of the normalized correlation coefficient of these templates at each point in a test image. To cope with scale variations, a set of templates at different scales was used.

The problems associated with the scale variations can be significantly reduced by using hierarchical correlation. For face recognition, the templates corresponding to the significant facial feature of the test images are compared in turn with the corresponding templates of all of the images in the database, returning a vector of matching scores computed through normalized cross correlation. The similarity scores of different features are integrated to obtain a global score that is used for recognition. Other similar methods that use correlation or higher order statistics revealed the accuracy of these methods but also their complexity.

To handle rotations, templates from different views are used. After the pose is determined, the task of recognition is reduced to the classical correlation method in which the facial feature templates are matched to the corresponding templates of the appropriate view based models using the cross correlation coefficient. However this computational approach is expensive, and it is sensitive to lighting conditions.

### 2.3.5 Matching pursuit-based methods

Many template-based face detection and recognition systems use a matching pursuit filter to obtain the face vectors. The matching pursuit algorithm applied to an image iteratively selects from a dictionary of basis functions the best decomposition of the image by minimizing the residue of the image in all iterations. The algorithm constructs the best decomposition of a set of images by iteratively optimizing a cost function, which is determined from the residues of the individual images. The dictionary of basis functions consists of two-dimensional wavelets, which give a better image representation than the PCA- and LDA-based techniques where the images were stored as vectors. For recognition the cost function is a measure of distances between faces and is maximized at each iteration. For detection the goal is to find a filter that clusters together in similar templates (the mean for example), and is minimized in each iteration. The feature represents the average value of the projection of the templates on the selected basis.

### 2.3.6 Support vector machine approach

Face recognition is a $K$ class problem, where $K$ is the number of known individuals. So, support vector machines (SVMs) which deal with binary classification methods can be applied to such a situation, by reformulating the face recognition problem and reinterpreting the output of the SVM classifier. The problem is formulated as a problem in difference space, which models dissimilarities between two facial images. In difference space, the face recognition is a two-class problem denoting dissimilarities between faces of the same person and dissimilarities between faces of different people. By modifying the interpretation of the decision surface generated by SVM, a similarity metric can be generated between faces that are learned from examples of differences between faces.

### 2.3.7　Selected works on face classifiers

Success of any face recognition system depends on the proper choice of classifiers. Among several classification algorithms, two simple and well-known decision rules are the $k$-nearest neighbour (k-NN) rule [73] and minimum distance classifier [74]. In face recognition, these classifiers are often combined with subspace algorithms. If prior probabilities and density functions are known, then Bayes classifier is one of the best performing classifiers in face recognition [74]. However, in many situations, the probability density functions are assumed to be multivariate normal, and their parameters are estimated from the given set of observations.

There also are some classifiers which try to minimize the error. In general, mean square error (MSE) between the classifier output and the target value is used. The neural network-based classifiers like multilayer perceptron and support vector machine [75] are also used. One of the well-known statistical approaches for obtaining a linear decision boundary between classes is Fisher's linear discriminant (FLD) function. Extraction of features using principal components, followed by FLD, showed significant improvement in recognition rates [76]. Nonlinear version of FLD, known as kernel Fisher discriminant (KFD), is also used for face classification [54], [77].

In recent years, the combined classifier approach showed better analytical results. A classifier combination is especially useful [78], if the individual classifiers are largely independent. Various re-sampling techniques like rotation and bootstrapping are also used. In case of real-time face recognition, however, the receiver operating characteristic (ROC) curve is used. A threshold with equal error rate (EER) is found to yield a decision rule that has been applied in face classification [79].

## 2.4　Face reconstruction techniques

Synthetic face generation for the purpose of face recognition has been explored in recent years. Two-dimensional (2D) to three-dimensional (3D) reconstruction and generation of new faces with various shapes and appearances have been successfully accomplished. 2D to 2D face reconstruction [80] and the active appearance model generated from the 2D face images are powerful methods. So far modeling is concerned with four main approaches: (a) active appearance models (AAMs) [80], (b) manifolds [81], (c) geometry-driven face synthesis methods [68] including face animation

[82] and (d) expression mapping techniques [83],[84]. 3D models include face mesh frames, morphable models and depth map-based models. These models need to incorporate high quality graphics and complex animation algorithms.

Flynn et al. [85] provided a survey of approaches and challenges in 3D and multi-modal 3D + 2D face recognition. 3D head poses are derived from 2D to 3D feature correspondences [86]. Face recognition based on fitting a 3D morphable model with statistical texture is also being undertaken [87]. Multi-view face reconstruction in 2D space is done by manifold analysis. Geometry-driven face synthesis [83] and reflectance models are also proposed [88]. Expression mapping techniques are found useful in 2D face generation. Table 2.4 provides comparative appraisal on advantages and disadvantages of some useful techniques of 2D face synthesis. Some significant advantages and disadvantages of 3D face synthesis are shown in Table 2.5.

### 2.4.1 Three-dimensional face recognition

A vast majority of face recognition research and commercial face recognition systems use typical intensity images of the face which are referred to as two-dimensional (2D) images. At no point in the recognition process is a three-dimensional (3D) model of a face constructed, nor do the algorithms make strong and explicit use of the fact that the images are the result of observing a 3D face. Unlike the 2D facial image, 3D facial surface is insensitive to illumination, head pose and cosmetics. Moreover, 3D data can be used to produce invariant measures out of the 2D data. While in 2D face recognition a conventional camera is used to produce a 2D face image, 3D face registration requires a more sophisticated sensor, capable of acquiring depth information, usually referred to as depth or range camera or 3D scanner. However, for recognition purposes, 3D shape of the face is usually acquired together with a 2D intensity image. In this case, the 2D image can be thought of as a texture map overlaid on the 3D shape. A range image, also sometimes called a depth image, is an image in which the pixel value reflects the distance from the sensor to the imaged surface. A range image, a shaded model and a wire-frame mesh are common alternatives for displaying 3D face data. In another approach a generic, morphable 3D face model is used as an intermediate step in matching two 2D images for face recognition. This approach does not involve the sensing or matching of 3D shape descriptions; instead, a 2D image is mapped onto a deformable 3D model, and the 3D model with texture is used to produce a set of synthetic 2D images for the matching process. Variations of this type of approach are already used in many commercial face recognition systems. Some 3D sample images are shown in Figure 2.4.

A typical 3D face recognition system is built from the following units: an image-acquisition and pre-processing unit, a feature-extraction unit and a similarity-measure and classification unit. In the subsequent sections, the units are presented in detail with examples of implementations that have emerged in the literature. The output of a common 3D sensor is a set of 3D points

**TABLE** 2.4: Comparative advantages/disadvantages of 2D to 2D face synthesis

| Researchers | Method | Advantages/ Disadvantages |
|---|---|---|
| Cootes et al. [89] | Uses AAM and ASM models. Statistical texture and shape analysis techniques are used. | Results are satisfactory. The initial feature points on face images are manually annotated in the training stage. |
| Huang et al. [81] | Faces in manifold are reconstructed with expressions. | Expressions are generated in neutral faces. Many face images taken from different angles are needed for the construction of a manifold |
| Liu et al. [68] | Expression ratio images are reconstructed using the ERI algorithm | Applications are done on 2D faces. The method can only map expressions present in the probe image. |
| Zhang et al. [83] | Facial expressions were synthesized using a feature point set | Expression editing software is available. An additional method is required for the generation of feature point set. |
| Pyun et al. [90] | In this expression mapping technique, geometry-controlled image warping method is used. | Morphing can be done easily since the knowledge of geometry of face is known. Geometrical properties of the face under consideration are to be found and stored. |
| Pighin et al. [91] | Basis expression space is created for every person in the training set. | Expressions of another person can be inherited. Construction of a person's expression space needs prior computations and also requires large memory space. |
| Neely et al. [92] | Morphing operators are used for the construction of new faces. | Easy to implement. Limited number of expressions are generated. Smooth transition from one face image to the other is not possible. |

**TABLE 2.5**: Comparative advantages/disadvantages of 3D face synthesis

| Researchers | Method | Advantages/Disadvantages |
| --- | --- | --- |
| Tao et al.[86] | 3D head poses designed | Direct method derived from 2D images. High computation complexity |
| Romdhani et al. [93] | 3D morphable models are used | Different poses and illuminations are added in morphable models. More computations are needed for identification |
| Blanz et al. [87] | Morphable models of 3D faces are created using laser scan | Depth information included in laser scan. Computational load is very high for the identification purpose |
| Blanz et al. [87], Flynn et al. [85] | 3D model derived from single 2D image | Single image required. Face geometrics and face texture database are used |
| Yin et al. [94] | Model generated from orthogonal views | Depth information is available. Orthogonal images used are difficult to get |

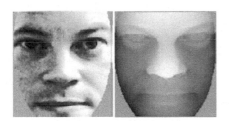

**FIGURE 2.4**: Typical 3D images used for face classification

of a scanned surface, with the values of the $x, y$ and $z$ components at each point. The 3D data is usually presented as a point cloud or a range image. The point cloud is a set of $(x; y; z)$ coordinates of scanned points from the object surface. A range image (or a depth image) can be obtained by the projection of scanned 3D points onto the $(x; y)$ plane. The range image is formatted in a similar way to a 2D intensity image, but with the difference that in the range image the pixel intensities are proportional to the depth components of a scanned object ($z$ coordinates).

In the acquired 3D image, the face detection and localization are usually performed first. Detection denotes a process where the presence and the number of faces in the image are determined. Assuming that the image contains only one face, the localization task is to find the location and size (and sometimes also the orientation) of a facial region. Most methods for face localization in 3D images are based on an analysis of the local curves of the facial surface. This gives a set of possible points for the locations of the characteristic facial parts, such as the location of the nose, eyes and mouth, through which the exact location, size and orientation of the facial area can be determined. Based on the locations of these points, the face area can be cut from the rest of the image and eventually re-scaled and rotated to the normal pose.

### 2.4.1.1 Feature extraction

As in 2D face detection and recognition, the purpose of feature extraction is to extract the compact information from the images that is relevant for distinguishing between the face images of different people and stable in terms of the photometric and geometric variations in the images. One or more feature vectors are extracted from the facial region. The existing feature-extraction methods can be divided into the groups of global and local operations.

### 2.4.1.2 Global feature extraction

In global feature extraction methods, feature vectors from the whole face region are extracted. The majority of the global 3D facial-feature-extraction methods have been derived from methods originally used on 2D facial images, where 2D gray-scale images are replaced by range images. Global-feature methods require the precise localization and normalization of the orientation, scale and illumination for robust recognition.

Principal component analysis (PCA) is the most widespread method for global feature extraction and is used for feature extraction from 2D face images and also from range images. Other popular global feature extraction methods, such as linear discriminant analysis (LDA) and independent component analysis (ICA) are also used on range images. The global features not only reduce the data dimensionality, but also retain the spatial relationship among the different parts of the face.

The use of global features is prevalent in face recognition systems based on images acquired in a controlled environment. In global feature-based recognition systems, localization and normalization are often performed by the manual labelling of characteristic points on the face. Automatic localization and normalization is generally achieved using the iterative closest-point algorithm (ICP).

Local feature extraction methods extract a set of feature vectors from a face, where each vector holds the characteristics of a particular facial region. The local features extraction methods have advantages over the global features in uncontrolled environments, where the variations in facial illumination, rotation, expressions and scale are present. The process of local feature extraction can be divided into two parts. In the first part, the interest points on the face region are detected. In the second part, the points of interest are used as locations at which the local feature vectors are calculated. The interest points can be detected as extrema in the scale-space, resulting in the invariance of features to the scale. The interest points can also be detected by analysing the curves of the face features and by the elastic bunch graph method, where the nodes of a rectangular grid cover the facial region.

### 2.4.1.3 Three-dimensional morphable model

The morphable model is a three-dimensional (3D) representation that enables the accurate modelling of any illumination and pose as well as the separation of these variations from the rest (identity and expression). It is a generative model consisting of a linear 3D shape and appearance model plus an imaging model, which maps the 3D surface onto an image. The 3D shape and appearance are modelled by taking linear combinations of a training set of example faces.

The main step of model construction is to build the correspondences of a set of 3D face scans. Given a single face image, the algorithm automatically estimates 3D shape, texture and all relevant 3D scene parameters like pose, illumination, etc. In the second step, the correspondences are computed between each of the scans and a reference face mesh. The registered face scans are then aligned such that they do not contain a global rigid transformation. Then a principal component analysis is performed to estimate the statistics of the 3D shape and color of the faces. The recognition task is achieved measuring the Mahalanobis distance between the shape and texture parameters of the models in the gallery and the fitting model.

# Chapter 3

## Subspace-based face recognition

## 3.1  Introduction

Human face images are usually represented by thousands of pixels encoded in high-dimensional array; however, they are intrinsically embedded in a very low-dimensional subspace. The use of subspace for representation of face images helps to reduce the so-called curse of dimensionality in subsequent classification. Dimension reduction of a dataset can be achieved by extracting the features, provided that the new features contain most of the information of the given dataset. To put it in a different way, the dimensionality reduction can be done if the mean squared error (MSE) or the sum of variances of the elements (which are going to be eliminated), are minimum. Subspace-based methods are such techniques of dimensionality reduction. Further the subspace representation helps in suppressing the variations of lighting conditions and facial expressions. Two of the most widely used subspace methods for face detection and face recognition are the principal component analysis (PCA) [95] and the Fishers linear discriminant analysis (LDA), though there are dozen of dimension reduction algorithms are available for selecting effective subspaces for the representation of face images [63].

## 3.2   Principal component analysis

The principal component analysis-based face recognition method was proposed in [5] and became very popular. Using this method a subset of principal directions (principal components) can be found in a set of the training faces. Then the face images are projected into the space of these principal components and the feature vectors are obtained. Face recognition is performed by comparing these feature vectors using different distance measures. Basically, in the PCA-based face recognition method the eigenvectors and eigenvalues of the covariance matrix of the training data are calculated. PCA produces the optimal linear least-squares decomposition of a training set. Because the basis vectors constructed by PCA had the same dimension as the input face images, they were named eigenfaces.

PCA representation of an image $X_o(m,n)$ of size $d_1 \times d_2$ pixels is interpreted as a point in $\Re^{d_1 \times d_2}$ space. If the mean image of the $N$ number of the lexicographic ordered training image $x_i$ is

$$\bar{m} = \sum_{i=1}^{N} x_i \qquad (3.1)$$

and the corresponding mean centered image is

$$w_i = x_i - \bar{m} \qquad (3.2)$$

then, a set of $e_i$'s can be obtained which has the largest possible projection onto each of the $w_i$'s. The objective is to find a set of $N$ orthonormal vectors $e_i$ for which the quantity

$$\lambda_i = \frac{1}{N} \sum_{k=1}^{N} (e_i^{\mathrm{T}} w_k)^2 \qquad (3.3)$$

is maximized with the orthonormal constraint given by

$$e_l^{\mathrm{T}} e_k = \delta_{lk} \qquad (3.4)$$

The eigenvectors $e_i$ and eigenvalues $\lambda_i$ of the covariance matrix $C = W^{\mathrm{T}}W$ are calculated, where $W$ is a matrix composed of the column vectors $w_i$ placed side by side. The weight vector of the trained face images is treated as a feature vector or face descriptor and is calculated as

$$v_i = e_i^{\mathrm{T}} w_i \qquad (3.5)$$

where $e_i$ is the eigenvector obtained through PCA and T stands for the transpose operation.

In Figure 3.1 the face feature extraction process using the PCA method is shown. Twelve training images of size $d_1 \times d_2$ are taken from the YaleB database with three differently illuminated faces from four different face classes. In Figure 3.1 the training faces are labelled with $x_i$, where $x_i$ represents the lexicographic version of the corresponding image $X_i$. In other words, it can be said that in reshaping the vector $x_i$ of length $d = d_1 \times d_2$ the image $X_i$ is formed of size $d_1 \times d_2$. The above statement is also applicable for $w_i$ and $e_i$. Since three lighting conditions are used, three different sets of illumination are obtained with respect to lighting angles such as (1) azimuth (A) $= -5^o$, elevation (E) $= -10^o$; (2) azimuth (A) $= +0^o$, elevation (E) $= +90^o$ and (3) azimuth (A) $= -85^o$, elevation (E) $= -20^o$. Each lighting subset contains four face images (i.e, four different persons). From twelve training faces mean face and mean subtracted faces are developed.

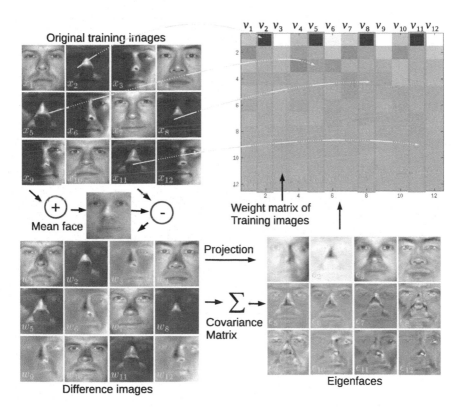

**FIGURE 3.1**: Block diagram of face feature extraction using principal component analysis

Eigenvalues $\lambda_i$ and corresponding eigenvectors $e_i$ are calculated from the covariance matrix. The associated eigenvalues allow the eigenvectors to characterize the variation among the face images. Figure 3.2 shows that the first eigenvector account for 30% of the variance in the dataset, while the first three eigenvectors together account for just over 75%, and with the first five eigenvectors, 90% of the variation in the dataset is reached. Increasing the number of eigenvectors generally increases recognition accuracy as the eigenvectors can be thought of as a set of features that accounts for the variation between face images. Each image location contributes to each eigenvector so that the eigenvectors can be displayed as a sort of ghostly faces termed eigenfaces [5]. Features of the training faces are extracted according to Equation 3.5 and the feature vectors $v_i$s are obtained. The image form of the weight matrix containing $v_i$s columnwise is shown in Figure 3.1. The dashed line in Figure 3.1 shows the feature vector $v_i$ corresponding to training images.

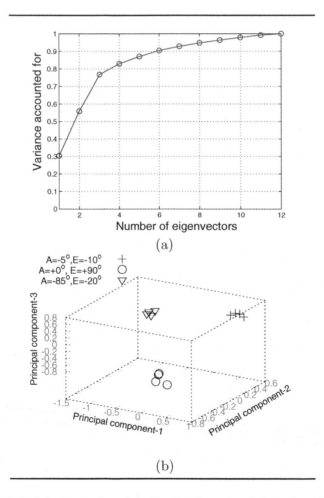

(a)

(b)

**FIGURE 3.2**: (a) Scree plot; (b) distribution of weights of twelve training faces in 3D space developed by three principal components

A scree plot is a graphical display of the variance of each component in the dataset which is used to determine how many components are retained in order to explain a high percentage of the variation in the data. It is drawn in Figure 3.2(a). An interesting conclusion can be drawn for the 3D plot shown in Figure 3.2(b); that instead of taking twelve eigenvectors, only the first three are taken corresponding to the first three principal components. Hence, for each training image only three weight values are obtained and those values are considered as a point in a three-dimensional space as plotted in Figure 3.2(b). From Figure 3.2(b) it may be seen that three different clusters are formed corresponding to three different lighting conditions. In other words, by using PCA, it is possible to represent the illumination variation of faces in the form of their weights. For the same type of illumination, however, different faces have almost the same weight values and so they are clustered as shown in Figure 3.2(b).

Therefore, mathematically, PCA maximizes the variance in the projected subspace for a given dimensionality, decorrelates the training face images in the projected subspace and maximizes the mutual information between the appearance of face images. One main limitation of the eigenface technique is that the class labels of face images cannot be explored in the process of learning the projection matrix for dimension reduction.

---

MATLAB code for eigenface for face recognition

---

```
% Principal Component Analysis for face recognition
% M training images, sized N pixels wide by N pixels tall
% c recognition images, also sized N by N pixels
% Mp = desired number of principal components

% Feature Extraction:
% merge column vector for each training face
X = [x1 x2 ... xm]
% compute the average face
me = mean(X,2)
A = X - [me me ... me]
% avoids N^2 by N^2 matrix computation of [V,D]=eig(A*transpose(A))
% only computes M columns of U: A=U*E*transpose(V)
[U,E,V] = svd(A,0)
eigVals = diag(E)
lmda = eigVals(1:Mp)
% pick face-space principal components (eigenfaces)
P = U(:,1:Mp)
% store weights of training data projected into eigenspace
train_wt = transpose(P)*A

Nearest-Neighbor Classification:
% A2 created from the recog data (in similar manner to A)
```

```
recog_wt = transpose(P)*A2
% euclidean distance for ith recog face, jth train face
euDis(i,j) = sqrt((recog_wt(:,j)-train_wt(:,i)).^2)
```

### 3.2.1   Two-dimensional principal component analysis

Two-dimensional principal component analysis (2DPCA) is based on 2D matrices rather than 1D vectors. That is, the image matrix does not need to be previously transformed into a vector. Instead, an image covariance matrix can be constructed directly using the original image matrices. In contrast to the covariance matrix of PCA, the size of the image covariance matrix using 2DPCA is much too small. The image covariance scatter matrix is given by

$$G_t = \frac{1}{M} \sum_{j=1}^{M} (A_j - \overline{A})(A_j - \overline{A})$$

The matrix $G_t$ is called the image covariance (scatter) matrix and can be evaluated directly using the training image samples.

Suppose that there are $M$ training image samples and the $j$-th training image is denoted by an $m \times n$ matrix $A_j$ where $A_j(j = 1, 2, \ldots, M)$ and the average image of all training samples is denoted by $\overline{A}$. In fact, the optimal projection axes, $X_1; \ldots; X_d$, are the orthonormal eigenvectors of $G_t$ corresponding to the first $d$ largest eigenvalues. It may be noted that each principal component of 2DPCA is a vector, whereas the principal component of PCA is a direction and therefore is a scalar.

---

MATLAB code for implementation of two-dimensional PCA

---

```
function [ WA,WB ] = pca2d( A,B,D )
%perform 2-dimensional PCA on training set A
of size mxnxN and test set B of size mxnxP
% i.e. there are N training and P test samples (images)
each of size mxn
      % D is the dimension to which A will be reduced
      A=double(A);B=double(B);
      [m n N]=size(A);[m n P]=size(B);
      total=A(:,:,1);
      for k=2:N
          total=total+A(:,:,k);
      end
      miu=total/N; % mean of A
      for k=1:N
          A(:,:,k)=A(:,:,k)-miu; % adjust A
      end
```

```
G=zeros(n,n);
for k=1:N
    G=G+transpose(A(:,:,k))*A(:,:,k);
end
G=G/N;
[y,l]=eig(G);% find eigen value and eigen vector
l=diag(l);
% find first D highest Eigen values
  and store the associated Eigen
% vectors in Y
[val,ind]=sort(l,"descend");
% sort Eigen values in descending order
Y=[];
for j=1:D
    Y=[Y y(:,ind(j))];
end
Y=Y./D; % normalize Y
for k=1:N
    X(:,:,k)=A(:,:,k)*Y;
end
WA=X
% find space projection projB of test set B
for k=1:P
    B(:,:,k)=B(:,:,k)-miu; % adjust A
end
for k=1:P
    WB(:,:,k)=B(:,:,k)*Y;
end
end
```

### 3.2.2 Kernel principal component analysis

In recent years, the reformulation of linear techniques using the kernel function has led to successful face recognition techniques. Kernel PCA (KPCA) is a reformulation of traditional linear PCA in a high-dimensional space and is constructed using a kernel function. It computes the principal eigenvectors of the kernel matrix, rather than those of the covariance matrix. The application of PCA in the kernel space provides kernel PCA the property of constructing nonlinear mappings. Because of increase in dimensionality, the mapping the huge data is made implicit (and economical) by the use of kernel functions which satisfy Mercer's theorem, given by

$$K = K(x, y) = (\phi(x), \phi(y)) = \phi(x)^T \phi(y)$$

where kernel evaluations in the input space correspond to dot products in the higher-dimensional feature space.

Since kernel PCA is a kernel-based method, the mapping performed by Kernel PCA relies on the choice of the kernel function. Possible choices for the kernel function include the linear kernel (making Kernel PCA equal to traditional PCA), the polynomial kernel such as $k(x, y) = (x^T y + 1)^2$ and the Gaussian kernel such as $k(x, y) = e^{\frac{-\|x-y\|^2}{2\sigma^2}}$ .

---

MATLAB code for implementation of Kernel PCA

---

```
function [ WA,WB ] = pcaKernel( A,B,D )}
    \texttt{%perform Kernel PCA on training set A and test set B
    % D is the dimension to which A will be reduced
        A=double(A);B=double(B);
[M N]=size(A);[M P]=size(B);
    miu=mean(transpose(transpose(A)));}
\texttt{% find row-wise mean of A
        for j=1:N
            A(:,j)=A(:,j)-miu; % adjust A
        end
        KA=((transpose(A)*A)+4).^2; % kernel of A
        Kmiu=mean(transpose(transpose(KA)));
        for j=1:N
            KA(:,j)=KA(:,j)-Kmiu; % adjust KA
        end
        oneA=ones(N,N)./N;
        KA=KA-oneA*KA-KA*oneA+oneA*KA*oneA;
        [y,l]=eig(KA/N);
% find eigen value and eigen vector
        l=diag(l);
        % find first D highest Eigen values
         and store the associated Eigen
        % vectors in Y
        [val,ind]=sort(l,"descend");
% sort Eigen values in descending order
        Y=[];
        D=D;
        for j=1:D
            Y=[Y y(:,ind(j))];
        end
        Y=Y./D; % normalize Y
        X=KA*Y;
        WA=X*transpose(KA);
% D-dimensional space projection of training images
    KB=((transpose(B)*A)+4).^2;
    oneB=ones(P,N)./N;
    KB=(KB-(oneB*KA)-(KB*oneA)+(oneB*KA*oneA));
    WB=X*transpose(KB);
end
```

## 3.3 Fisher linear discriminant analysis

PCA finds a linear projection of high-dimensional data into a lower dimensional subspace such that the variance retained is maximized and the least square reconstruction error is minimized. However the direction of maximum variance may be useless for classification as the components identified by a PCA do not necessarily contain any discriminative information at all. Therefore, the projected samples are smeared together and a classification becomes impossible. For example, when substantial changes in illumination and expression are present, much of the variation in the data is due to these changes. The PCA techniques essentially select a subspace that retains most of that variation, and consequently the similarity in the face subspace is not necessarily determined by the identity.

To address difficulties of the PCA, another representative subspace method for face recognition known as Fisherface is attempted. In contrast to eigenface, Fisherface finds class-specific linear subspaces. The dimension reduction algorithm used in Fisherface is Fishers linear discriminant analysis (FLDA), which simultaneously maximizes the between-class scatter and minimizes the within-class scatter of the face data. In general, Fisherface outperforms eigenface due to the utilized discriminative information.

Figure 3.3 to Figure 3.8 describe the shortcomings of PCA during classification. Two sets of two-dimensional data belonging to class-1 and class-2 are plotted in Figure 3.3 and the mean subtracted data plot is shown in Figure 3.4. Figure 3.4 also shows two orthogonal eigenvectors corresponding to the covariance matrix of the datasets. Eigenvector-1 shows the direction of the maximum variance of the datasets.

Class-1 data=([1.,2.],[2.,3.],[3.,3.],[4.,5.],[5.,5.])
Class-2 data=([1.,0.],[2.,1.],[3.,1.],[3.,2.],[5.,3.],[6.,5.])

Figure 3.5 shows the data reconstruction while both eigenvectors are used and the original data are perfectly reconstructed. Data are linearly separable with this reconstruction. Reconstruction error is visible when only one eigenvector (corresponding to the maximum variance) is exploited as shown in Figure 3.6.

It is interesting to observe from Figure 3.7 and Figure 3.8 that the projected data in reduced space (1D from 2D) are non-linearly separable while the original datasets are linearly separable. From this point of view it may be concluded that PCA fails to classify these data as the direction of the principal component is along the maximum variance. If the class discriminant information is retained in the variance of the datasets, PCA works well. However, in the example, the class discriminatory information is retained in their respective mean and hence PCA fails to separate data linearly.

In order to find the combination of features that separates best between classes the linear discriminant analysis maximizes the ratio of between-

*Subspace-based face recognition*

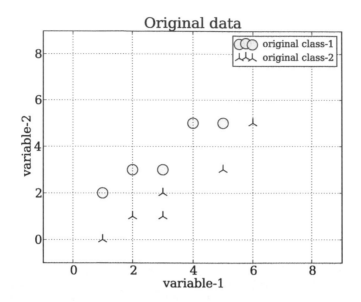

**FIGURE 3.3**: Two-dimensional data of two different classes

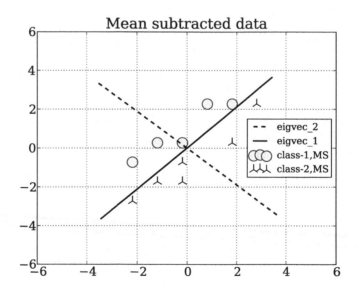

**FIGURE 3.4**: Mean subtracted (MS in figure) data of two different classes

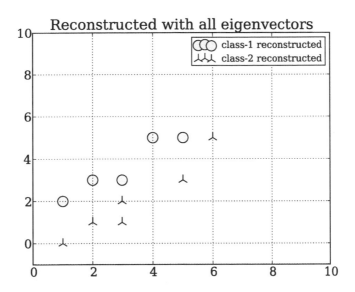

**FIGURE 3.5**: Data reconstruction with all eigenvectors

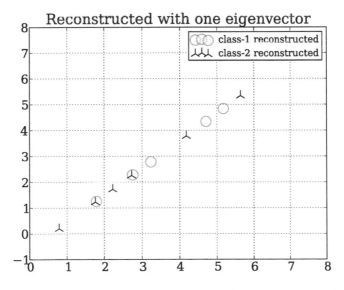

**FIGURE 3.6**: Data reconstruction with one eigenvector corresponding to maximum variance. Some reconstruction error is present

*Subspace-based face recognition*

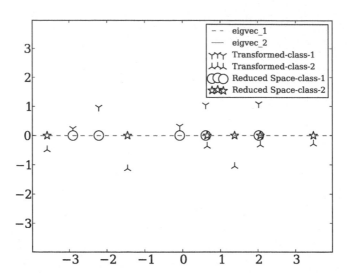

**FIGURE 3.7**: Data in transformed space where the coordinate system is based on eigenvectors

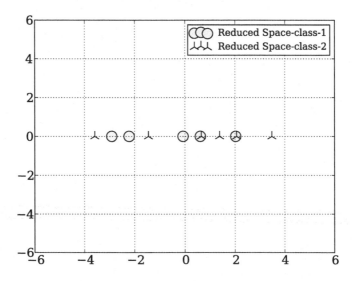

**FIGURE 3.8**: Nonlinearly separable data in reduced space

classes to within-classes scatter. The idea is based on a simple property that same classes should cluster tightly together, while different classes are highly separated from each other [96]. LDA thus seeks to reduce dimensionality while preserving as much of the class discriminatory information as possible.

---

MATLAB code for implementation of Fisherface

---

```
% Fisherface

%% same training & recognition images, also sized N by N pixels
% P1 = eigenface result
% Feature Extraction:
% same as eigenface
A = X - [me me ... me]
% compute N^2 by N^2 between-class scatter matrix
for i=1:c
    Sb = Sb + clsMeani*transpose(clsMeani)

% compute N^2 by N^2 within-class scatter matrix
for i=1:c, j=1:ci
    Sw = Sw + (X(j)-clsMeani)*transpose(X(j)-clsMeani)
% project into (N-c) by (N-c) subspace using PCA
Sbb = transpose(P1)*Sb*P1
Sww = transpose(P1)*Sw*P1
% generalized eigenvalue decomposition
% solves Sbb*V = Sww*V*D
[V,D] = eig(Sbb,Sww)
eigVals = diag(D)
lmda = eigVals(1:Mp)
P = P1*V(:,1:Mp)

% store training weights
train_wt = transpose(P)*A
%% Nearest-Neighbor Classification:
% same as eigenface
```

---

## 3.3.1 Fisher linear discriminant analysis for two-class case

Consider d-dimensional samples $x_1, x_2, ..., x_N$, of which $N_1$ belong to $c_1$-class and $N_2$ belong to $c_2$-class, where $N = N_1 + N_2$, a scalar $y$ is obtained by projecting the samples $x$ onto a line (C-1 space, here, C = 2), that is,

$$y = w^T x \tag{3.6}$$

where $w$ is a d-dimensional vector as $x$ which provides the direction of the one that maximizes the separability of the scalars among all of the possible lines.

In order to find a good projection vector, a measure of separation between the projections is defined. The distance between the projected means can then be selected as an objective function. Projected mean of class $c_1$ is given as

$$\mu_{p1} = \frac{1}{N_1} \sum_{i=1}^{N_1} w^T x_i = w^T \mu_1 \qquad (3.7)$$

where

$$\mu_1 = \frac{1}{N_1} \sum_{i=1}^{N_1} x_i \qquad (3.8)$$

Similarly,

$$\mu_{p2} = w^T \mu_2 \qquad (3.9)$$

The distance between the projected means $J(w)$ is given by

$$J(w) = \|\mu_{p1} - \mu_{p2}\| = \|w^T(\mu_{p1} - \mu_{p2})\| \qquad (3.10)$$

However, the distance between the projected means is not a very good measure, since it does not take into account the standard deviation within the classes. Figure 3.9 depicts the disadvantage of taking only the difference of the projected means. The axis (large variance) with the projected means has greater distance and provides poor class separability. In contrast, better class separability along the axis (small variance) is achieved when the distance between the projected means is less.

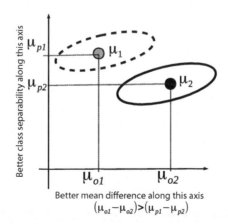

**FIGURE 3.9**: Better difference in projected mean gives poor separation where better separation along the axis does not provide large difference in projected means

One solution to this problem is proposed by Fisher by maximizing a function that represents the difference between the projected means, normalized by a measure of the within-class variability. This is sometimes called scatter, which is proportional to variance. Scatter measures the spread of data around the mean just like variance. If $y_i = w^T x_i$ is the projected sample, scatter for the projected samples of class $c_1$ is given by

$$s_{p1}^2 = \sum_{i=1}^{N_1} (y_i - \mu_{p1})^2, \qquad y_i \in c_1 \tag{3.11}$$

and for class $c_2$ is

$$s_{p2}^2 = \sum_{i=1}^{N_2} (y_i - \mu_{p2})^2, \qquad y_i \in c_2 \tag{3.12}$$

Thus a Fisher linear discriminant projects on line in the direction $w$ and maximizes

$$J(w) = \frac{(\mu_{p1} - \mu_{p2})^2}{s_{p1}^2 + s_{p2}^2} \tag{3.13}$$

Maximization of $J(w)$ in Equation 3.13 is achieved when the numerator term $(\mu_{p1} - \mu_{p2})^2$ is maximized while the denominator term, the within-class scatters, $s_{p1}^2 + s_{p2}^2$ is minimized. It is desirable that the projected class means should be as far away as possible, where the samples of respective classes cluster around the projected means. Figure 3.10 illustrates this.

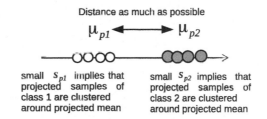

Distance as much as possible

small $s_{p1}$ implies that projected samples of class 1 are clustered around projected mean

small $s_{p2}$ implies that projected samples of class 2 are clustered around projected mean

**FIGURE 3.10**: Physical interpretation of maximization of $J(w)$

In order to find the optimum projection $w^o$, $J(w)$ is expressed as an explicit function of $w$. The scatter of the projected samples for class $c_1$ can then be rewritten as

$$
\begin{aligned}
s_{p1}^2 &= \sum_{i=1}^{N_1} (y_i - \mu_{p1})^2, \qquad y_i \in c_1 \\
&= \sum_{i=1}^{N_1} (w^T x_i - w^T \mu_1)^2 \\
&= \sum_{i=1}^{N_1} w^T (x_i - \mu_1)(x_i - \mu_1)^T w \\
&= w^T S_1 w \tag{3.14}
\end{aligned}
$$

where

$$S_1 = \sum_{i=1}^{N_1} (x_i - \mu_1)(x_i - \mu_1)^T, \qquad x_i \in c_1 \tag{3.15}$$

represents the within-class scatter for class $c_1$. Similarly,

$$s_{p2}^2 = w^T S_2 w \tag{3.16}$$

where

$$S_2 = \sum_{i=1}^{N_2} (x_i - \mu_2)(x_i - \mu_2)^T, \qquad x_i \in c_2 \tag{3.17}$$

represents the within-class scatter for class $c_2$.

Hence, total within-class scatter of the projected samples $s_{p1}^2 + s_{p2}^2$ can be represented by within-class scatter before projection as

$$s_{p1}^2 + s_{p2}^2 = w^T S_1 w + w^T S_2 w = w^T S_W w \tag{3.18}$$

where $S_W$ is the total within-class scatter matrix before projection or within-class scatter matrix for original space.

Similarly, the difference between the projected means can be expressed in terms of the means in the original feature space as

$$
\begin{aligned}
(\mu_{p1} - \mu_{p2})^2 &= (w^T \mu_1 - w^T \mu_2)^2 \\
&= w^T (\mu_1 - \mu_2)(\mu_1 - \mu_2)^T w \\
&= w^T S_B w
\end{aligned}
\tag{3.19}
$$

The matrix $S_B$ is called the between-class scatter. Since $S_B$ is the outer product of two vectors, its rank is at most one.

Finally Fisher criterion can be expressed in terms of $S_W$ and $S_B$ as

$$J(w) = \frac{w^T S_B w}{w^T S_W w} \tag{3.20}$$

Hence, $J(w)$ is a measure of the difference between class means (encoded in the between-class scatter matrix) normalized by a measure of the within-class scatter matrix. To find the maximum of $J(w)$, it is differentiated and is equated to zero, as given by

$$
\begin{aligned}
\frac{d}{dw} J(w) &= \frac{d}{dw} \left( \frac{w^T S_B w}{w^T S_W w} \right) \\
&= (w^T S_W) 2 S_B w - (w^T S_B w) 2 S_W w \\
&= (w^T S_W w)/(w^T S_W w) 2 S_B w - \frac{w^T S_B w}{w^T S_W w} 2 S_W w \\
&= 0
\end{aligned}
\tag{3.21}
$$

or

$$S_B w \;=\; J(w) S_W w$$
$$S_W^{-1} S_B w \;=\; J(w) w \tag{3.22}$$

which is nothing but a generalized eigenvalue problem.

The optimum direction $w^o$ can be obtained by evaluating the eigenvector of $S_W^{-1} S_B$ corresponding to the largest eigenvalue. This is known as Fishers linear discriminant, although it is not a discriminant but rather a specific choice of direction for the projection of the data down to one dimension.

Figure 3.11 and Figure 3.12 describe the condition of better class separation where FLDA is applied in contrast to Figure 3.7 and Figure 3.8 where PCA is applied on the same set of data.

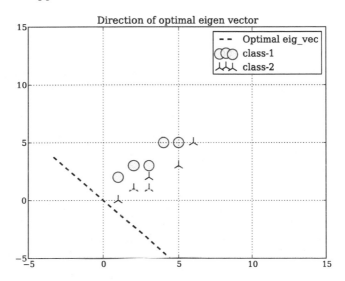

**FIGURE 3.11**: The direction of optimal eigenvector when the same set of data has been used (as in case of PCA)

Although FLDA has shown promising performance on face recognition, it has a few limitations. FLDA discards the discriminative information preserved in covariance matrices of different classes and therefore it cannot find a proper projection for subsequent classification when samples are taken from complex distributions, except those from Gaussian distributions. Moreover, FLDA tends to merge classes which are close together in the original feature space and when the size of the training set is smaller than the dimension of the feature space, FLDA has an undersampled problem.

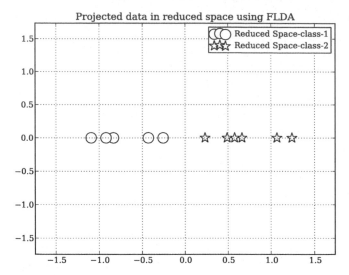

**FIGURE 3.12**: Better classification is obtained using FLDA because in the reduced space, the data are linearly separable

## 3.4  Independent component analysis

While PCA minimizes the sample covariance of the data, independent component analysis (ICA) minimizes higher-order dependencies as well, and the components found by ICA are designed to be non-Gaussian [97], [98]. ICA yields a linear projection like PCA but with different properties; that is, approximate reconstruction, non-orthogonality of the basis, and the near-factorization of the joint distribution into marginal distributions of the non-Gaussian independent components. The 2D subspace recovered by ICA appears to reflect the distribution of the data much better than the subspace obtained with PCA.

PYTHON code for implementation of principal component analysis

```
# Classification for two class case using PCA

import numpy as np
from matplotlib import pyplot as plt
from operator import itemgetter

plt.rc("font", family="serif",size=18,weight="light")
#plt.rc("text",usetex=True)
```

```
#class1 = np.array([[2.5,2.4],[2.2,2.9],[3.1,3.0],[2.3,2.7],[1.9,2.2]])
#class2 = np.array([[0.5,0.7],[1,1.1],[1.5,1.6],[1.1,0.9],[2,1.6]])
plt.close("all")
class1 =np.array([[1.,2.],[2.,3.],[3.,3.],[4.,5.],[5.,5.]])
N1 = len(class1)
class2 = np.array([[1.,0.],[2.,1.],[3.,1.],[3.,2.],[5.,3.],[6.,5.]])
N2 = len(class2)
data = np.vstack((class1,class2))

plt.figure(1)
plt.scatter(data[0:N1,0],data[0:N1,1],s=240,c=[0.9,0.9,0.9],\
marker="o",label="original class-1",alpha=0.9)
plt.scatter(data[N1:N1+N2,0],data[N1:N1+N2,1],\
s=240,c=[0.0,0.0,0.0],marker="4",label="original class-2")
plt.grid(axis="both")
plt.legend(loc=0,prop={"size":14})
plt.title("Original data")
plt.xlim(-1,1.5*data.max())
plt.ylim(-1,1.5*data.max())
plt.xlabel("variable-1")
plt.ylabel("variable-2")
plt.show()
m = np.array([data.mean(axis=0)])
M = np.tile(m,(data.shape[0],1))
D = data - M
Cov = np.cov(D.T)
CovMat = float(1./(D.shape[0]-1.)) * np.dot(D.T,D)
val,vec = np.linalg.eig(CovMat)
tmp = np.zeros((val.shape))
tmpvec = np.zeros((vec.shape))
for i in range(len(val)):
    a = max(enumerate(val), key=itemgetter(1))[0]
    tmp[i] = val[a]
    tmpvec[:,i] = vec[:,a]
    val[a]=0

plt.figure(2)
plt.scatter(D[0:N1,0],D[0:N1,1],s=240,c=[0.9,0.9,0.9],\
marker="o",label="class-1,MS",alpha=0.9)
plt.scatter(D[N1:N1+N2,0],D[N1:N1+N2,1],s=240,c=[0,0,0],\
marker="4",label="class-2,MS")
plt.grid(axis="both")
plt.plot([-5*tmpvec[0,0],5*tmpvec[0,0]] ,[-5*tmpvec[1,0],\
5*tmpvec[1,0]],"--k",lw=2,label="eigvec_1")
plt.plot([-5*tmpvec[0,1],5*tmpvec[0,1]] ,[-5*tmpvec[1,1],\
5*tmpvec[1,1]],"-k",lw=2,label="eigvec_2")
plt.xlim(-data.max(),data.max())
plt.ylim(-data.max(),data.max())
plt.legend(loc=0,prop={"size":14})
```

```
plt.title("Mean subtracted data")
plt.show()
## Data reconstruction with all eigen vectors
transData = np.dot(D,tmpvec) # taking all eigen vectors

plt.figure(3)
pc = tmpvec
reconstructed = np.dot(transData,pc.T) + M
plt.scatter(reconstructed[0:N1,0],reconstructed[0:N1,1],\
s=240,c=[0.9,0.9,0.9],marker="o",label="class-1 reconstructed")
plt.scatter(reconstructed[N1:N1+N2,0],\
reconstructed[N1:N1+N2,1],s=240,c=[0.0,0.0,0.0],\
marker="4",label="class-2 reconstructed")
plt.grid(axis="both")
plt.legend(loc=0,prop={"size":14})
plt.xlim(0,10)
plt.ylim(-1,10)
plt.title("Reconstructed with all eigenvectors")
plt.show()
## Data reconstruction with  eigen vector having maximum variance
plt.figure(4)
pc = np.array([tmpvec[:,0]]).T
rec = np.dot(D,pc) + M
plt.scatter(rec[0:N1,0],rec[0:N1,1],s=240,c=[0.9,0.9,0.9],\
marker="o",label="class-1 reconstructed",alpha=0.5)
plt.scatter(rec[N1:N1+N2,0],rec[N1:N1+N2,1],s=240,c=[0.0,0.0,0.0],\
marker="4",label="class-2 reconstructed")
plt.plot([-10*tmpvec[0,0],10*tmpvec[0,0]] ,[-10*tmpvec[1,0],\
10*tmpvec[1,0]],"--k",lw=2,label="eigvec_1")
plt.grid(axis="both")
plt.legend(loc=0,prop={"size":14})
plt.xlim(0,8)
plt.ylim(-1,8)
plt.title("Reconstructed with one eigenvector")
plt.show()
plt.figure(5)
rec = rec -M
rec[:,1] = 0
plt.scatter(rec[0:N1,0],rec[0:N1,1],s=200,c=[1,1,1],\
marker="o",label="Reduced Space-class-1")
plt.scatter(rec[N1:N1+N2,0],rec[N1:N1+N2,1],s=160,c=[1,1,1],\
marker="*",label="Reduced Space-class-2")
plt.xlim(-(rec.max()+0.5),rec.max()+0.5)
plt.ylim(-(rec.max()+0.5),rec.max()+0.5)
plt.grid(axis="both")
plt.legend(loc=0,prop={"size":12})
plt.show()
```

PYTHON code for implementation of Fisher linear discriminant analysis

```python
# Classification for two class case using FLDA
import numpy as np
from matplotlib import pyplot as plt
from operator import itemgetter
plt.rc("font", family="serif",size=12,weight="light")
plt.close("all")
class1 =np.array([[1.,2.],[2.,3.],[3.,3.],[4.,5.],[5.,5.]])
#class1 =np.array([[4,2],[2,4],[2,3],[3,6],[4,4]])
N1 = len(class1)
class2 = np.array([[1.,0.],[2.,1.],[3.,1.],[3.,2.],[5.,3.],[6.,5.]])
#class2 = np.array([[9,10],[6,8],[9,5],[8,7],[10,8]])
N2 = len(class2)

m1 = np.array([class1.mean(axis=0)])
m2 = np.array([class2.mean(axis=0)])
d1 = class1 - np.tile(m1,(class1.shape[0],1))
d2 = class2 - np.tile(m2,(class2.shape[0],1))
S1 = float(1./(class1.shape[0]-1.)) * np.dot(d1.T,d1)
S2 = float(1./(class2.shape[0]-1)) * np.dot(d2.T,d2)
Sw = S1 + S2
Sb = np.dot((m1-m2).T,(m1-m2))

invSw = np.linalg.inv(Sw)
invSwSb = np.dot( invSw , Sb)
val,vec = np.linalg.eig(invSwSb)

tmp = np.zeros((val.shape))
tmpvec = np.zeros((vec.shape))
for i in range(len(val)):
    a = max(enumerate(val), key=itemgetter(1))[0]
    tmp[i] = val[a]
    tmpvec[:,i] = vec[:,a]
    val[a]=0
w = np.array([tmpvec[:,0]]).T
#""""" eigenvector with largest eigenvalue """""
#plt.plot([-5*vec[0,1],5*vec[0,1]] ,[-5*vec[1,1],5*vec[1,1]],
"-k",lw=2,label="eigvec_1")
plt.figure(1)
plt.scatter(class1[:,0],class1[:,1],s=240,c=[0.9,0.9,0.9],\
marker="o",label="class-1",alpha=0.9)
plt.scatter(class2[:,0],class2[:,1],s=240,c=[0.,0.,0.],\
marker="4",label="class-2")
plt.xlim(-5,15)
plt.ylim(-5,15)
plt.plot([-5*w[0],20*w[0]] ,[-5*w[1],20*w[1]],\
"--k",lw=2,label="Optimal eig_vec")
```

```
plt.legend(loc=0,prop={"size":14})
plt.grid(axis="both")
plt.title("Direction of optimal eigen vector")
plt.show()

p1 = np.dot(class1,tmpvec) # projection of class-1 on optimal eigenvector
p2 = np.dot(class2,tmpvec)
p1[:,1] = 0 # taking the projected values only on optimal vector
p2[:,1] = 0
rec = np.vstack((p1,p2))
plt.figure(2)
plt.scatter(rec[0:N1,0],rec[0:N1,1],s=250,c=[1,1,1],\
marker="o",label="Reduced Space-class-1")
plt.scatter(rec[N1:N1+N2,0],rec[N1:N1+N2,1],s=250,\
c=[1,1,1],marker="*",label="Reduced Space-class-2")
plt.xlim(-(rec.max()+0.5),rec.max()+0.5)
plt.ylim(-(rec.max()+0.5),rec.max()+0.5)
plt.grid(axis="both")
plt.legend(loc=0,prop={"size":12})
plt.title("Projected data in reduced space using FLDA")
plt.show()
```

# Chapter 4

# Face detection by Bayesian approach

## 4.1    Introduction

The term Bayesian refers to the eighteenth century mathematician and theologian Thomas Bayes, who provided the first mathematical treatment of a non-trivial problem of Bayesian inference. The technique is basically based on matching of images for the purpose of detection using a probabilistic measure of similarity. The performance advantage of this technique over the Euclidean distance measure used in PCA has been established by the fact the measure does not exploit knowledge of critical variations. Detection of faces in both gray and color scene can be performed using the Bayes decision rule. Multiple face detection can also be performed using Bayesian approach.

## 4.2    Bayes decision rule for classification

Bayesian decision theory is a fundamental statistical approach to the problem of classification. This approach is based on quantifying the trade-offs between various classification decisions using probability and the costs that accompany such decisions. It makes the assumption that the decision problem is posed in probabilistic terms, and that all of the relevant probability values are known. It permits also to determine the optimal (Bayes) classifier against which all other classifiers can be compared and sometimes it also helps to predict the errors.

Let us consider a hypothetical problem of designing a classifier to separate two classes of faces in a randomly arranged face database. While searching the face images at random it is difficult to predict which class of face will emerge next as the faces in the database are randomly arranged, but it can safely be said that the image from either of the classes would appear next. Let $w$ denote the state of the face database, with $w = w_1$ for one class (class A) of face and $w = w_2$ for the other class (class B) of face. Because the state of the face database is unpredictable, $w$ is considered as a variable that must be described probabilistically. At this point it is assumed that there is some prior probability $P(w_1)$ with which the next face that may be accessed is from class A and similarly prior probability $P(w_2)$ is associated with class B. These prior probabilities reflect prior knowledge of accessing an arbitrary face image of the database. For a decision on the face class that might be accessed next, the value of the prior probabilities will decide for $w_1$ if $P(w_1) > P(w_2)$; otherwise decide for $w_2$. This rule makes sense if only one database is considered having two classes of face images. The same rule needs to be repeatedly applied, if the database contains many classes of face images which may result in large errors. The probability of error for this type of decision is

$$P(error) = \min\{P(w_1), P(w_2)\}$$

In general no decisions are taken with such little information (prior probabilities), and an estimation measurement of random variable $x$ is exploited for better classification. The distribution of random variable $x$ depends on the state of class $w_1$ and is denoted as $p(x|w_1)$. This estimation of $x$ for the class $w_1$ is termed the class conditional probability function or likelihood estimation. This likelihood estimation can be expressed as a Gaussian (normal) distribution.

### 4.2.1 Gaussian distribution

The most important probability distribution for continuous variables is called the normal or Gaussian distribution. For single real-valued variable $x$, the Gaussian distribution is defined by

$$N(x|\mu, \sigma^2) = \frac{1}{(2\pi\sigma^2)^{1/2}} exp\left\{-\frac{1}{2\sigma^2}(x - \mu)^2\right\} \tag{4.1}$$

which is governed by two parameters, the mean $\mu$ and the variance $\sigma^2$.

Figure 4.1 shows two 1D Gaussian distributions of two datasets $X = [59, 61, 48, 45, 67, 55]$ and $Y = [29, 19, 17, 30, 22, 25]$.

The square root of the variance is called the standard deviation $\sigma$. For

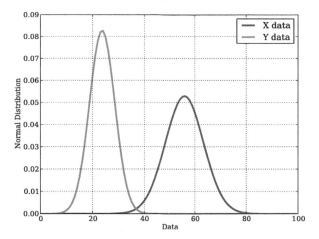

**FIGURE 4.1**: Two 1D Gaussian distributions of different mean and standard deviation

a valid probability density, the Gaussian distribution satisfies the equation given by

$$\int_{-\infty}^{\infty} N(x \mid \mu, \sigma^2) dx = 1 \tag{4.2}$$

and

$$N(x \mid \mu, \sigma^2) > 0 \tag{4.3}$$

The average value of $x$ under the Gaussian distribution is given by

$$E[x] = \int_{-\infty}^{\infty} N(x \mid \mu, \sigma^2) x \, dx = \mu \tag{4.4}$$

and the second order moment

$$E[x^2] = \int_{-\infty}^{\infty} N(x \mid \mu, \sigma^2) x^2 \, dx = \mu^2 + \sigma^2 \tag{4.5}$$

From the above two equations, the variance can be calculated as,

$$var[x] = E[x^2] - E[x]^2 = \sigma^2 \tag{4.6}$$

Gaussian distribution defined over a $D$-dimensional vector $\mathbf{x}$ of continuous variables is given by

$$N(\mathbf{x} \mid \boldsymbol{\mu}, \boldsymbol{\Sigma}) = \frac{1}{(2\pi)^{D/2} |\boldsymbol{\Sigma}|^{1/2}} exp\left\{ -\frac{1}{2} (\mathbf{x} - \boldsymbol{\mu})^T \boldsymbol{\Sigma}^{-1} (\mathbf{x} - \boldsymbol{\mu}) \right\} \tag{4.7}$$

where $\boldsymbol{\mu}$ is a $D$-dimensional mean vector, $\boldsymbol{\Sigma}$ is a $D \times D$ covariance matrix and $|\boldsymbol{\Sigma}|$ represents the determinants of the covariance matrix.

Figure (4.2) shows a bivariate Gaussian distribution which is completely described by two parameters, $\boldsymbol{\mu}$ and $\boldsymbol{\Sigma}$.

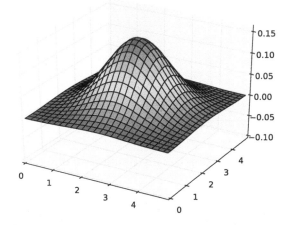

**FIGURE 4.2**: Bivariate Gaussian distribution

The mean and covariance matrix for a two dimensional random vector $\mathbf{X} = \begin{pmatrix} \mathbf{x}_1 \\ \mathbf{x}_2 \end{pmatrix}$ are

$$\boldsymbol{\mu} = E\begin{pmatrix} \mathbf{x}_1 \\ \mathbf{x}_2 \end{pmatrix} = \begin{pmatrix} \mu_1 \\ \mu_2 \end{pmatrix}$$

and

$$\boldsymbol{\Sigma} = \begin{pmatrix} var(\mathbf{x}_1) & covar(\mathbf{x}_1, \mathbf{x}_2) \\ covar(\mathbf{x}_1, \mathbf{x}_2) & var(\mathbf{x}_2) \end{pmatrix} = \begin{pmatrix} \sigma_1^2 & \rho\sigma_1\sigma_2 \\ \rho\sigma_1\sigma_2 & \sigma_2^2 \end{pmatrix}$$

The joint probability density function (pdf) can now be written for the two-dimensional form. Two dimensions represent two features, one feature corresponds to $\mathbf{x}_1$ and the other feature corresponds to $\mathbf{x}_2$. These are given as

$$N(\mathbf{x} \mid \boldsymbol{\mu}, \boldsymbol{\Sigma}) = \frac{1}{(2\pi)|\boldsymbol{\Sigma}|^{1/2}} exp\left\{-\frac{1}{2}(\mathbf{x} - \boldsymbol{\mu})^T \boldsymbol{\Sigma}^{-1}(\mathbf{x} - \boldsymbol{\mu})\right\} \qquad (4.8)$$

The determinant of covariance matrix is calculated as

$$|\boldsymbol{\Sigma}| = (1 - \rho^2)\sigma_1^2\sigma_2^2$$

This further helps to write the Mahalanobis distance as

$$\left\{-\frac{1}{2}(\mathbf{x} - \boldsymbol{\mu})^T \boldsymbol{\Sigma}^{-1}(\mathbf{x} - \boldsymbol{\mu})\right\} =$$

$$\frac{-1}{2(1 - \rho^2)}\left(\frac{(x_1 - \mu_1)^2}{\sigma_1^2} - 2\rho\frac{(x_1 - \mu_1)(x_2 - \mu_2)}{\sigma_1\sigma_2} + \frac{(x_2 - \mu_2)^2}{\sigma_2^2}\right) \qquad (4.9)$$

Hence, the joint pdf of the two-dimensional random variable $\mathbf{X}$ is given by

$$N(\mathbf{X} \mid \boldsymbol{\mu}, \boldsymbol{\Sigma}) = \frac{1}{2\pi\sigma_1\sigma_2\sqrt{1-\rho^2}} \times$$

$$exp\left\{\frac{-1}{2(1-\rho^2)}\left(\frac{(x_1-\mu_1)^2}{\sigma_1^2} - 2\rho\frac{(x_1-\mu_1)(x_2-\mu_2)}{\sigma_1\sigma_2} + \frac{(x_2-\mu_2)^2}{\sigma_2^2}\right)\right\} \tag{4.10}$$

If $\rho = 0$, the above equation can be written as

$$N(\mathbf{X} \mid \boldsymbol{\mu}, \boldsymbol{\Sigma}) = \frac{1}{2\pi\sigma_1\sigma_2} \times$$

$$exp\left\{\frac{-1}{2}\left(\frac{(x_1-\mu_1)^2}{\sigma_1^2} + \frac{(x_2-\mu_2)^2}{\sigma_2^2}\right)\right\} \tag{4.11}$$

The above equation is a product of the marginal distributions of $\mathbf{x}_1$ and $\mathbf{x}_2$ and $\mathbf{x}_1, \mathbf{x}_2$ which are considered as independent.

For different types of covariance matrices, the Gaussian distributions are different as shown in Figure (4.3). Consider that a data set of observations

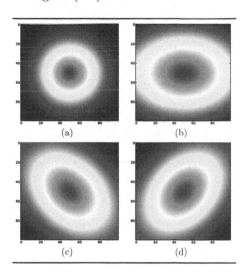

(a)      (b)

(c)      (d)

**FIGURE 4.3**: Different Gaussian distributions: (a) spherical Gaussian, (b) diagonal covariance Gaussian, (c) full covariance Gaussian with negative non-diagonals and (d) full covariance Gaussian with positive non-diagonals

$\mathbf{x} = (x_1, ..., x_N)^T$ represents $N$ observations of a scalar variable $x$. Data points that are drawn independently from the Gaussian distribution are said to be independent and identically distributed. The joint probability of two independent events is given by the product of marginal probabilities for each event separately. For such a data set $\mathbf{x}$, the probability of the data set can be written in the form of an equation, given $\mu$ and $\sigma^2$.

$$p(\mathbf{x} \mid \mu, \sigma) = \prod_{i=1}^{N} N(\mathbf{x}_i | \mu, \sigma) \tag{4.12}$$

PYTHON code for 1D Gaussian function

```python
# 1D Gaussian function

import numpy as np
from matplotlib import pyplot as plt

plt.rc("font", family="serif",size=12,weight="light")
plt.close("all")

X = np.array([59,61,48,45,67,55])
Y = np.array([29,19,17,30,22,25])
mx = X.mean(axis=0)
sx = np.std(X)
Nx = np.zeros((100,1),dtype=float)
my = Y.mean(axis=0)
sy = np.std(Y)
Ny = np.zeros((100,1),dtype=float)

for i in range(0,100):
    Nx[i,:] = (1/np.sqrt(2*np.pi*sx**2)) * \
    np.exp(-(0.5/sx**2)*(i-mx)**2)\

    Ny[i,:] = (1/np.sqrt(2*np.pi*sy**2)) * np.exp(-(0.5/sy**2)*(i-my)**2)
plt.Figure  (1)
plt.xlabel("Data")
plt.ylabel("Normal Distribution")
plt.plot(Nx,"-b",label="X data",linewidth=3)
plt.plot(Ny,"-r",label="Y data",linewidth=3)
plt.legend(loc="upper right")
plt.grid(axis="both")
plt.show()
```

PYTHON code for bivariate Gaussian function

```python
# Bivariate Gaussian function

from numpy import *
import numpy as np
from mpl_toolkits.mplot3d import axes3d
```

```
import matplotlib.pyplot as plt
from matplotlib import cm

plt.close("all")

K1 = array([[1,0.9],[0.3,1]])
m1 = array([2.5,2.5])
normfac = 1/(pow(np.linalg.det(K1),0.5)*(2*pi))
x = arange(0,5,0.05,dtype=float)
p = zeros((size(x,0),size(x,0)),dtype=float)
for i in range(0,size(x,0)):
    for j in range(0,size(x,0)):
        fst = dot(array([[x[i],x[j]] - m1]) , (np.linalg.inv(K1)))
        p[i,j] = normfac * exp(-0.5
        * dot (fst ,array([[x[i],x[j]] -
        m1]).T))

xx,yy = meshgrid(x,x)
fig = plt.Figure  ()
ax = Figure    gca(projection="3d")
ax.plot_surface(xx,yy,p,rstride=4,cstride=4,linewidth=1\
,antialiased=False,cmap=cm.coolwarm)
#cset = ax.contour(xx, yy, p, offset=-0.1, cmap=cm.coolwarm)
#ax.view_init(elev=65., azim=-110.)
ax.set_zlim(-0.1, p.max())
#fig, ax = plt.subplots()
#cset = ax.contour(x,x,p)

fig, ax = plt.subplots()
h=ax.contour(p, cmap=cm.RdBu, vmin=abs(p).min(),\
 vmax=abs(p).max(), extent=[0, 6, 0, 6])
numsamp = 1000
[lamda,eta] = np.linalg.eig(K1)
coeffs = np.random.rand(numsamp,2)
samples = dot(coeffs,eta.T) + np.ones((numsamp,1))*m1
plt.plot(samples[:,0],samples[:,1],".k")
plt.show()
```

---

### 4.2.2 Bayes theorem

Suppose both the prior probabilities $P(w_1)$, $P(w_2)$ and class-conditional densities $p(x \mid w_1), p(x \mid w_2)$ are known. The joint probability density of finding a pattern existing in the category $w_j, \{j = 1, 2\}$ and that has feature value $x$ can be written as

$$p(w_j, x) = P(w_j \mid x)p(x) = p(x \mid w_j)P(w_j) \tag{4.13}$$

Rearranging the above equation, the famous Bayes rule is given by

$$P(w_j \mid x) = \frac{p(x \mid w_j)P(w_j)}{p(x)} \tag{4.14}$$

In the case of two categories,

$$p(x) = \sum_{j=1}^{2} p(x \mid w_j)P(w_j) \tag{4.15}$$

Equation 4.14 can be written informally as

$$posterior = \frac{likelihood \times prior}{evidence} \tag{4.16}$$

Baye's formula shows that by observing the value of $x$, the prior probability $P(w_j)$ can be converted to the posterior probability $P(w_j|x)$. The probability of the state is $w_j$ and that feature value $x$ has been measured. $p(x|w_j)$ is called the likelihood of $w_j$ with respect to $x$, a term chosen to indicate that, other things being equal, the category $w_j$, for which $p(x|w_j)$ is large, is more likely to be the true category. Notice that it is the product of the likelihood and the prior probability that is most important in determining the posterior probability. The so-called evidence factor $p(x)$ can be viewed as a scale factor that guarantees that the sum of posterior probabilities is one.

The decision after observing $x'$ can be evaluated from Baye's formula as

$$Decide \begin{cases} w_1; & if \quad P(w_1|x) > P(w_2|x) \\ w_2; & if \quad P(w_2|x) > P(w_1|x) \end{cases}$$

### 4.2.3   Bayesian   decision   boundaries   and   discriminant   function

The most important issue, in the design of a pattern classifier, is the placement of a decision boundary which separates the classes. In the case of placing the Bayesian decision boundary, initially it is that the boundary partitions the input space into two regions, $R_1$ and $R_2$. Then the probability $P_{error}$ of a feature $x$ being assigned to the wrong class is set as

$$\begin{aligned} P_{error} &= P(x \in R_2, w_1) + P(x \in R_1, w_2) \\ &= \int_{-\infty}^{x_d} p(x|w_2)P(w_2)dx + \int_{-\infty}^{x_d} p(x|w_1)P(w_1)dx \end{aligned} \tag{4.17}$$

where $x_d$ is the point of placement of the decision boundary.

In order to minimize $P_{error}$, the selection of the decision boundary for $R_1$ and $R_2$ is governed under the constraints:

- point $x$ lies in $R_1$, for decision of class $w_1$, if $p(x|w_1)P(w_1) > p(x|w_2)P(w_2)$

- point $x$ lies in $R_2$, for decision of class $w_2$, if $p(x|w_1)P(w_1) < p(x|w_2)P(w_2)$

An optimal Bayes's classification chooses the class with maximum posterior probability $P(w_j|x)$ using the discriminant function given as

$$g_j(\mathbf{x}) = ln p(\mathbf{x}|w_j) + ln P(w_j) \qquad (4.18)$$

In case of a two-class decision problem, the class discriminant function for classes $w_1$ and $w_2$, respectively, is given as

$$g_1(\mathbf{x}) = -\frac{1}{2}(\mathbf{x} - \boldsymbol{\mu}_1)^T \boldsymbol{\Sigma}_1^{-1}(\mathbf{x} - \boldsymbol{\mu}_1) - ln|\boldsymbol{\Sigma}_1| - \frac{n}{2}ln 2\pi + ln P(w_1) \qquad (4.19)$$

and

$$g_2(\mathbf{x}) = -\frac{1}{2}(\mathbf{x} - \boldsymbol{\mu}_2)^T \boldsymbol{\Sigma}_2^{-1}(\mathbf{x} - \boldsymbol{\mu}_2) - ln|\boldsymbol{\Sigma}_2| - \frac{n}{2}ln 2\pi + ln P(w_2) \qquad (4.20)$$

If $g_1(\mathbf{x}) > g_2(\mathbf{x})$, $\mathbf{x}$ is assigned to $w_1$ and if $g_1(\mathbf{x}) < g_2(\mathbf{x})$, $\mathbf{x}$ is assigned to $w_2$. This decision can be implemented in a more compact form by defining an alternative discriminant function $g(\mathbf{x})$ given by

$$g(\mathbf{x}) = g_2(\mathbf{x}) - g_1(\mathbf{x}) \qquad (4.21)$$

Then the decision rule is implemented in the following way: $\mathbf{x}$ is assigned to $w_1$ if $g(\mathbf{x}) < 0$ and $\mathbf{x}$ is assigned to $w_2$ if $g(\mathbf{x}) > 0$. The expression for $g(\mathbf{x})$ is

$$g(\mathbf{x}) = -\frac{1}{2}(\mathbf{x} - \boldsymbol{\mu}_1)^T \boldsymbol{\Sigma}_1^{-1}(\mathbf{x} - \boldsymbol{\mu}_1) - \frac{1}{2}(\mathbf{x} - \boldsymbol{\mu}_2)^T \boldsymbol{\Sigma}_2^{-1}(\mathbf{x} - \boldsymbol{\mu}_2) \qquad (4.22)$$

$$+\frac{1}{2}ln \frac{|\boldsymbol{\Sigma}_1|}{\boldsymbol{\Sigma}_2} - ln \frac{P(w_1)}{P(w_2)} \qquad (4.23)$$

The boundary of the decision rule can be found by setting $g(\mathbf{x}) = 0$. The boundary is elliptic as $\boldsymbol{\Sigma}_1 \neq \boldsymbol{\Sigma}_2$. Clearly if $\boldsymbol{\Sigma}_1 = \boldsymbol{\Sigma}_2$, the discriminant becomes linear.

---

MATLAB code for Bayes' boundary for linearly separable data

---

```
%% Bayes boundary for linearly separable data

x=linspace(-10,10,300);
y =linspace(-10,10,300);
[X,Y] = meshgrid(x,y);
mu = [3 6];
Sigma = [0.5 0; 0 2]; R = chol(Sigma);
```

```
z = repmat(mu,300,1) + randn(300,2)*R;
plot(z(:,1),z(:,2),"^r","MarkerSize",6);hold on
mu = [3 -2];
Sigma = [2 0; 0 2]; R = chol(Sigma);
z = repmat(mu,300,1) + randn(300,2)*R;
plot(z(:,1),z(:,2),"ob","MarkerSize",6);hold on
axis([-4 10 -10 10])
x=linspace(-10,10,100);
y = 3.335-1.125.*x+0.1875*x.^2;
plot(x,y,"-k","linewidth",3);
legend("class-1","class-2","Boundary");
```

---

MATLAB code for Bayes' boundary for non linearly separable data

---

```
%% Bayes boundary for non linearly separable data

syms x1 x2;
mu1 =[2,2];
sigma1 = [1,0;0,1];
mu2 =[3,-3];
sigma2=[1,0;0,1];
R1 = chol(sigma1);
z1 = repmat(mu1,300,1) + randn(300,2)*R1;
R2 = chol(sigma2);
z2 = repmat(mu2,300,1) + randn(300,2)*R2;
plot(z1(:,1),z1(:,2),"^g","MarkerSize",6);hold on
plot(z2(:,1),z2(:,2),"om","MarkerSize",6);hold on
g1 = -0.5*transpose([x1;x2]-transpose(mu1))*...
     inv(sigma1)*([x1;x2]-transpose(mu1))...
    -log(2*pi)-0.5*log(det(sigma1));
g2 = -0.5*transpose([x1;x2]-transpose(mu2))*...
     inv(sigma2)*([x1;x2]-transpose(mu2))...
    -log(2*pi)-0.5*log(det(sigma2));
g = g1-g2;
h = ezplot(g,[[-2,8],[-5,8]]);
set(h,"linewidth",3)
```

---

### 4.2.4   Density estimation using eigenspace decomposition

Density estimation using eigenspace decomposition is an unsupervised technique used for automatic detection of face or facial parts detection [99]. Instead of applying density estimation on the high dimensional space of the face image, by using the principal component analysis (PCA), the dimension is reduced.

For a given set of training images $\mathbf{x}_i, i = \{1, 2, \cdots, N_T\}$, where $N_T$ is the total number of training images for a given class $w$, the class membership or likelihood function $P(\mathbf{x} \mid w)$ is estimated. This estimation of density function is based on Gaussian distribution as described in previous sections. A training set of vectors $\{\mathbf{x}_i\}_{i=1}^{N_T}$, where $\mathbf{x} \in \Re^{mn}$, can be formed from a set of $m \times n$ images $\{\mathbf{X}_i\}_{i=1}^{N_T}$ by lexicographic ordering.

The basis functions of the Karhunen-Loeve transform are obtained by solving the eigenvalue problem given by

$$\Lambda = \phi^T \Sigma \phi$$

where $\Sigma$ is the covariance matrix of the data, $\phi$ are the eigenvectors and $\Lambda$ is the diagonal matrix of eigenvalues.

In PCA as stated in Chapter 3, the eigenvectors corresponding to the largest eigenvalues are identified. The principal component feature vector is obtained as

$$\mathbf{y} = \phi_M^T \mathbf{d}$$

where

$$\mathbf{d} = \mathbf{x} - \mathbf{m}$$

$\mathbf{m} \in \Re^{mn}$ is the mean of data and $\phi_M$ is the sub-matrix formed from $\phi$ taking $M$ eigenvectors corresponding to $M$ largest eigenvalues.

This corresponds to an orthogonal decomposition of vector space $\Re^N$ into two mutually exclusive and complementary subspaces, (i) features space or principal subspace $P = \{\phi_i\}_{i=1}^M$ containing principal components and (ii) its orthogonal complement $\bar{P} = \{\phi_i\}_{i=M+1}^N$, as shown in Figure 4.4. In partial

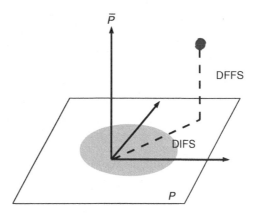

**FIGURE 4.4**: Principal subspace $P$ and orthogonal subspace $\bar{P}$ for Gaussian density

Karhunen-Loeve expansion, the residual reconstruction error is defined as

$$\epsilon^2(\mathbf{x}) = \sum_{i=M+1}^N y_i^2 = |\mathbf{d}|^2 - \sum_{i=1}^M y_i^2 \tag{4.24}$$

The equation can be computed from the first $M$ principal components and the $L_2$ norm of the mean subtracted images $\mathbf{d}$. The $L_2$ norm of every element $\mathbf{x} \in \Re^{N=mn}$ can also be decomposed in terms of its projections in two subspaces denoted as (i) distance-from-feature-space (DFFS) and (ii) distance-in-feature-space (DIFS).

The components of $\mathbf{x}$ that lie in the feature space can be interpreted in terms of the probability distribution of $y$ in $P$. The likelihood of an input pattern $\mathbf{x}$ under the Gaussian distribution with parameters $\mathbf{m}$ and $\Sigma$ obtained from the training patterns $\mathbf{x}_i, i = \{1, 2, \cdots, N_T\}$ is given by

$$P(\mathbf{x}|w) = \frac{1}{(2\pi)^{N/2}|\Sigma|^{1/2}} exp\left\{-\frac{1}{2}(\mathbf{x} - \mathbf{m})^T \Sigma^{-1}(\mathbf{x} - \mathbf{m})\right\} \qquad (4.25)$$

The Mahalanobis distance which is a sufficient statistic for characterizing this likelihood is calculated as

$$D(\mathbf{x}) = \mathbf{d}^T \Sigma^{-1} \mathbf{d} \qquad (4.26)$$

Using the eigenvectors and eigenvalues of $\Sigma$, the $\Sigma^{-1}$, $D(\mathbf{x})$ can be written as

$$
\begin{aligned}
D(\mathbf{x}) &= \mathbf{d}^T \Sigma^{-1} \mathbf{d} \\
&= \mathbf{d}^T [\phi \Lambda^{-1} \phi^T] \mathbf{d} \\
&= \mathbf{d}^T \phi \Lambda^{-1} \phi^T \mathbf{d} \\
&= \mathbf{y}^T \Lambda^{-1} \mathbf{y}
\end{aligned}
\qquad (4.27)
$$

Due to the diagonal nature, the Mahalanobis distance is reformulated as

$$D(\mathbf{x}) = \sum_{i=1}^{N} \frac{y_i^2}{\lambda_i}$$

Estimating $D(\mathbf{x})$ using only $M$ principal projections, the estimator $D(\mathbf{x})$ is given as

$$\hat{D}(\mathbf{x}) = \sum_{i=1}^{M} \frac{y_i^2}{\lambda_i} + \frac{1}{\rho}\left[\sum_{i=M+1}^{N} y_i^2\right] \qquad (4.28)$$

The likelihood estimate based on $\hat{D}(\mathbf{x})$ can be reformulated as the product of two marginal and independent Gaussian densities and is given by

$$
\begin{aligned}
\hat{P}(\mathbf{x}|w) &= \left\{\frac{exp(-0.5 \sum_{i=1}^{M} \frac{y_i^2}{\lambda_i})}{(2\pi)^{M/2} \prod_{i=1}^{M} \lambda_i^{0.5}}\right\} \times \left\{\frac{exp(-\frac{\epsilon^2(\mathbf{x})}{2\rho})}{(2\pi\rho)^{(N-M)/2}}\right\} \\
&= P(\mathbf{x}|w)\bar{P}(\mathbf{x}|w)
\end{aligned}
\qquad (4.29)
$$

where $P(\mathbf{x}|w)$ is the true marginal density in DIFS and $\bar{P}(\mathbf{x}|w)$ is the estimated marginal density in the orthogonal component DFFS.

## 4.3  Bayesian discriminant feature method

Chengjun Liu in [100] has established a Bayesian discriminating feature for frontal face detection by integrating the discriminating feature analysis technique with the statistical modelling and Bayes classifier. Feature analysis derives a discriminating feature vector by combining the input image, its 1D Haar wavelet representation and its amplitude projections. The Haar wavelets produce an effective representation for object detection; the amplitude projections capture the vertical symmetric distributions and the horizontal characteristics of face images. Statistical modelling estimates the conditional probability density functions of the face and non-face classes. The face class and non-face class is modelled as a multivariate normal distribution. Finally, the Bayes classifier is applied on the estimated conditional pdf to detect multiple frontal faces in a scene.

The Haar wavelet representation is effective for the human face in a scene. Haar wavelets are natural sets of basis functions which encode differences in average intensities between different regions of ian mage. Two types of 2D non-standard Haar wavelets are shown in Figure 4.5.

The amplitude projections are able to capture the vertical symmetric distributions and the horizontal characteristics of human face images. The

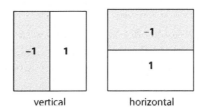

vertical          horizontal

**FIGURE 4.5**: 2D Non-standard vertical and horizontal Harr wavelets

1D Haar representation of an input image $I(i,j) \in \Re^{m \times n}$ gives two images as $I_h(i,j) \in \Re^{(m-1) \times n}$ and $I_v(i,j) \in \Re^{m \times (n-1)}$. Lexicographic ordering of these two images $I_h(i,j)$ and $I_v(i,j)$ generates two vectors $\mathbf{x}_h \in \Re^{(m-1)n \times 1}$ and $\mathbf{x}_v \in \Re^{(n-1)m \times 1}$, respectively. Horizontal and vertical difference images can be obtained as

$$I_h(i,j) = I(i+1,j) - I(i,j), \quad 1 \le i \le m, 1 \le j \le n \qquad (4.30)$$

$$I_v(i,j) = I(i,j+1) - I(i,j), \quad 1 \le i \le m, 1 \le j \le n \qquad (4.31)$$

$I_h(i,j)$ and $I_v(i,j)$ representations of a face and non-face image are shown in the second and third column of Figure 4.6.

The amplitude projections of $I(i,j)$ along its rows and columns form the

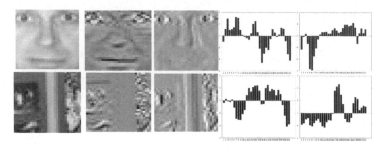

**FIGURE 4.6**: Top row represents the vertical, horizontal, row-wise amplitude projection and columnwise amplitude projection of a face image. Bottom row represents similar representations for non-face image

horizontal (row) and vertical (column) projections; $\mathbf{x}_r \in \Re^m$ and $\mathbf{x}_c \in \Re^n$, respectively, are given by

$$\mathbf{x}_r(i) = \sum_{j=1}^{n} I(i,j), \quad 1 \leq i \leq m \tag{4.32}$$

and

$$\mathbf{x}_c(i) = \sum_{i=1}^{m} I(i,j), \quad 1 \leq j \leq n \tag{4.33}$$

Both $\mathbf{x}_r \in \Re^m$ and $\mathbf{x}_c \in \Re^n$ are shown in bar-plot in the fourth and fifth column of Figure 4.6. The vectors $\mathbf{x}, \mathbf{x}_h, \mathbf{x}_v, \mathbf{x}_r, \mathbf{x}_c$ are normalized by subtracting from their respective mean and standard deviations to form a new feature vector $\hat{\mathbf{y}} = \{\mathbf{x}, \mathbf{x}_h, \mathbf{x}_v, \mathbf{x}_r, \mathbf{x}_c\}^T$, where $\mathbf{y} \in \Re^{N=3mn}$. Feature vector $\hat{\mathbf{y}}$ is again normalized to $\mathbf{y}$ as

$$\mathbf{y} = \frac{\hat{\mathbf{y}} - m}{\sigma} \tag{4.34}$$

where $m$ and $\sigma$ are the mean and standard deviation of the components of $\hat{\mathbf{y}}$.

### 4.3.1   Modelling of face and non-face pattern

The conditional density function of face class $w_f$ is modelled by the multivariate Gaussian distribution as

$$p(\mathbf{y}|w_f) = \frac{1}{(2\pi)^{N/2}|\mathbf{\Sigma}_f|^{1/2}} exp\left\{-\frac{1}{2}(\mathbf{y} - \mathbf{m}_f)^T\mathbf{\Sigma}_f^{-1}(\mathbf{y} - \mathbf{m}_f)\right\} \tag{4.35}$$

where $\mathbf{m}_f \in \Re^N$ and $\mathbf{\Sigma}_f \in \Re^{N \times N}$ are mean and covariance matrix of the face class. Taking the natural logarithm, the probability density function (PDF) is in the form

$$ln(p(\mathbf{y}|w_f) = -0.5\{(\mathbf{y} - \mathbf{m}_f)^T\mathbf{\Sigma}_f^{-1}(\mathbf{y} - \mathbf{m}_f)\} + Nln(2\pi) + ln|\mathbf{\Sigma}_f|\} \tag{4.36}$$

Using PCA, the covariance matrix $\Sigma_f$ can be factorized into

$$\Sigma_f = \phi_f \Lambda_f \phi_f^T$$

with

$$\phi_f \phi_f^T = \phi_f^T \phi_f = \mathbf{I}_N$$

and

$$\Lambda_f = diag\{\lambda_1, \lambda_2, \cdots, \lambda_N\}$$

where $\phi_f \in \Re^{N \times N}$ is an orthogonal matrix, $\Lambda_f \in \Re^{N \times N}$ is a diagonal eigenvalue matrix with eigenvalues in decreasing order ($\lambda_1 > \lambda_2 > ... \geq \lambda_N$) along the diagonal matrix and $\mathbf{I} \in \Re^{N \times N}$ is an identity matrix.

The principal components are established by the following vector $\mathbf{f} \in \Re^N$ given as

$$\mathbf{f} = \phi^T(\mathbf{y} - \mathbf{m}_f) \tag{4.37}$$

Hence, pdf can be written as

$$ln(p(\mathbf{y}|w_f)) = -0.5\left\{\mathbf{f}^T \Lambda_f^{-1} \mathbf{f} + Nln(2\pi) + ln|\Lambda_f|\right\} \tag{4.38}$$

The components of $\mathbf{f}$ are the ($M < N$) principal components.

From the discussion given in the previous section, the estimate of the remaining ($N - M$) eigenvalues can be represented as

$$\rho = \frac{1}{N - M} \sum_{k=M+1}^{N} \lambda_k \tag{4.39}$$

Previous derivations of pdf can now be rewritten as

$$ln(p(\mathbf{y}|w_f)) = -0.5\left\{\sum_{i=1}^{M} \frac{f_i^2}{\lambda_i} + \frac{|\mathbf{f} - \mathbf{m}_f|^2 - \sum_{i=1}^{M} f_i^2}{\rho} + \right.$$
$$\left. ln\left(\prod_{i=1}^{M} \lambda_i\right) + (N - M)ln(\rho) + Nln(2\pi)\right\} \tag{4.40}$$

where $f_i$s are the components of $\mathbf{f}$ defined in Equation 4.40.

Equation 4.40 states that the conditional density function of a face class can be estimated using,first, the $M$ principal components, input image, the mean face image and the eigenvalues of the face class.

Similarly the conditional density function can be modelled under the Gaussian distribution for non-face class, provided the non-face class patterns are very close to face class patterns. However, non-face class modelling needs sub-images from any natural image which does not contain any face. A set of equations established for face class can be written for the non-face class by suitably changing the notations.

## 4.3.2   Bayes classification using BDF

After modelling the conditional PDFs of the face and non-face classes, the Bayes classifier is utilized for multiple frontal face detection. The Bayes classifier yields the minimum error when the underlying PDFs are known. Bayes error is the optimal measure for feature effectiveness for classification, since it is a measure of class separability. The posterior probabilities of face class $w_f$ and non-face class $w_n$ for given $\mathbf{y}$ of any sub-image pattern of a scene, are $P(w_f|\mathbf{y})$ and $P(w_n|\mathbf{y})$ respectively. The pattern is classified to the face class or the non-face class according to the Bayes decision rule for minimum error given as

$$\mathbf{y} \in \left\{ \begin{array}{ll} w_f & if P(w_f|\mathbf{y}) > P(w_n|\mathbf{y}) \\ w_n & otherwise \end{array} \right. \tag{4.41}$$

These posterior probabilities $P(w_f|\mathbf{y})$ and $P(w_n|\mathbf{y})$ can be computed from conditional density functions and can be written for the face class as

$$P(w_f|\mathbf{y}) = \frac{p(\mathbf{y}|w_f)P(w_f)}{P(\mathbf{y})} \tag{4.42}$$

and for the non-face class as

$$P(w_n|\mathbf{y}) = \frac{p(\mathbf{y}|w_n)P(w_n)}{P(\mathbf{y})} \tag{4.43}$$

where $P(\mathbf{y})$ is the mixture density function

$$P(\mathbf{y}) = p(\mathbf{y}|w_n)P(w_n) + p(\mathbf{y}|w_f)P(w_f)$$

and $P(w_f), P(w_n)$ are the prior probabilities of face and non-face classes, respectively.

---

MATLAB code for face detection using BDF

---

```
%% Matlab Code for Face Detection using BDF
for i = 1:100
    img = im2double( imread(strcat("/home/pkb/scale/facepart/",...
        num2str(i),".jpg")));
    Y = bdffeature(img);
    Yf(:,i) = Y;
    clear Y
end
for i = 1:100
    img = im2double( imread(strcat("/home/pkb/scale/Nonface",...
        num2str(i),".jpg")));
    Y = bdffeature(img);
    Yn(:,i) = Y;
    clear Y
end
```

```
M = 60;N = size(Yf,2);
[vecf,valf,Mf] = bdfpca(Yf,M);
[vecn,valn,Mn] = bdfpca(Yn,M);
facecnt = 1;
for k = 1:350
    img = im2double( imread(strcat("/home/pkb/scale/Nonface/",...
        num2str(k),".jpg")));
    Y = bdffeature(img);
    Z = transpose(vecf)*(Y-Mf);
    zisq = Z.^2;
    lamda = valf(1:M);
    frac = zisq./lamda;
    fst =  sum(frac);
    ro = (1/(N-M))*sum(valf(M+1:N));
    snd = (norm(Y-Mf) - sum(zisq))/ro;
    trd = log(prod(valf(1:M)));
    frth = (N-M)*log(ro);
    deltf = (fst+snd+trd+frth)*10^-8+0.1;
    U = transpose(vecn)*(Y-Mn);
    uisq = U.^2;
    lamda = valn(1:M);
    frac = uisq./lamda;
    fst =  sum(frac);
    ep = (1/(N-M))*sum(valn(M+1:N));
    snd = (norm(Y-Mn) - sum(uisq))/ep;
    trd = log(prod(valn(1:M)));
    frth = (N-M)*log(ep);
    deltn = (fst+snd+trd+frth)*10^-7;
    if (deltf   < deltn) && (deltf<0)
        %display "face";
        facecnt
        facecnt=facecnt+1;
    else
        %display "nonface";
    end
end
```

---

MATLAB code for feature selection function using BDF

---

```
%% Feature selection function using BDF
function [Y,Xh] = bdffeature(img)
F=img(:);
F = (F-mean(F))/std(F);
for k = 1:size(img,1)-1
    h(:,k) =img(:,k+1)- img(:,k);
end
Xh = h(:);
```

```
Xh = (Xh-mean(Xh))/std(Xh);
for k = 1:size(img,2)-1
    v(k,:) =img(k+1,:)- img(k,:);
end
Xv = v(:);
Xv = (Xv-mean(Xv))/std(Xv);
Xr = sum(img,1);
Xr = (Xr-mean(Xr))/std(Xr);
Xc = sum(img,2);
Xc = (Xc-mean(Xc))/std(Xc);
Y = cat(1,F,Xh,Xv,transpose(Xr),Xc);
Y = (Y-mean(Y))/std(Y);
```

## 4.4    Experiments and results

To conduct the validity of the Bayesian method for face detection, training of both face and non-face data is needed. A set of face and non-face class images of size $32 \times 32$ are used for training. Figure 4.7 illustrates face and non face images used for training purposes.

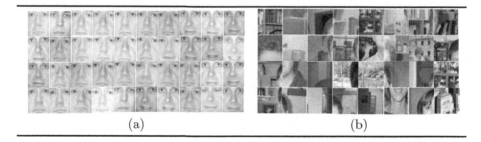

(a)                       (b)

**FIGURE 4.7**: Example (a) face images, (b) non-face images used for training

With these training images the Bayesian discriminating features are obtained and, subsequently, from these data, eigenspace decomposition is performed for the reduction of the dimensionality of **y**. From eigenspace decomposition, eigenvalues of both face and non-face classes are obtained along with $\mathbf{m}_f$ and $\mathbf{m}_n$.

For testing, an overlapping sub-images of size $32 \times 32$ is used and its BDFs is calculated. Eigen decomposition is performed over the test BDF and posterior probabilities are calculated for the test pattern using the Equations 4.40. The position of maximum posterior related to face class indicates the position of face in the target image. Some example face detection results are shown in Figure 4.8.

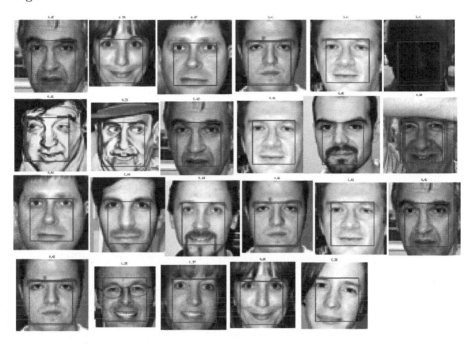

**FIGURE 4.8**: Face part detection in cropped frontal faces

---

MATLAB code for face detection using BDF

---

```
%% Matlab Code for Face Detection using BDF
for i = 1:99
    img = im2double( imread(strcat...
        ("/home/pradipta/PRADIPTA/Database/facetrain/",...
        num2str(i),".jpg")));
    Y = bdffeature(img);
    Yf(:,i) = Y;
    clear Y img
end
for i = 1:355
    img = im2double( imread(strcat...
        ("/home/pradipta/PRADIPTA/Database/Nonface/",...
        num2str(i),".jpg")));
    Y = bdffeature(img);
    Yn(:,i) = Y;
    clear Y
end
M = 15
```

```
Nf = size(Yf,2);
Nn = size(Yn,2);
[vecf,valf,Mf] = bdfpca(Yf,M);
[vecn,valn,Mn] = bdfpca(Yn,M);
facecnt = 1;
for siz = 0.3:0.01:0.48
    test = imread...
        ("home/pradipta/PRADIPTA/Database/CroppedFaces/147.jpg");
    if size(test,3)>1
        test = rgb2gray(test);
    end
    test = imresize(test,siz);
    figure
    imshow(test,[]);title(num2str(siz))
    for i = 1:size(test,1)-32
        for k=1:size(test,2)-32
            t = im2double(test(i:i+31,k:k+31));
            %t = histeq(t);
            Y = bdffeature(t);
            Z = transpose(vecf)*(Y-Mf);
            zisq = Z.^2;
            lamda = valf(1:M);
            frac = zisq./lamda;
            fst =  sum(frac);
            ro = (1/(Nf-M))*sum(valf(M+1:Nf));
            snd = (norm(Y-Mf) - sum(zisq))/ro;
            trd = log(prod(valf(1:M)));
            frth = (Nf-M)*log(ro);
            deltf = abs((fst+snd+trd+frth)*10^-7);
            U = transpose(vecn)*(Y-Mn);
            uisq = U.^2;
            lamda = valn(1:M);
            frac = uisq./lamda;
            fst =  sum(frac);
            ep = (1/(Nn-M))*sum(valn(M+1:Nn));
            snd = (norm(Y-Mn) - sum(uisq))/ep;
            trd = log(prod(valn(1:M)));
            frth = (Nn-M)*log(ep);
            deltn = abs((fst+snd+trd+frth)*10^-7);

            if (deltf   > deltn+1.5)
                rectangle("Position",[k i 32 32],...
                    "LineWidth",3, "EdgeColor","b");
            end
        end
    end
    clear test
end
```

# Chapter 5

## Face detection in color and infrared images

## 5.1   Introduction

The human visual system can distinguish hundreds of thousands of different color shades and intensities, but only around 100 shades of gray. Therefore, in a face image, a great deal of information is contained in the color, and the color information can then be used to simplify face detection and identification tasks. Although color appears to be a salient attribute of faces, past research has suggested that it confers little recognition advantage for identifying people. However, the performance of face recognition systems varies significantly according to the environment where face images are taken and according to the way user-defined parameters are adjusted. Recognition achieved on images taken in the visual spectrum remains limited particularly in outdoor environments and at low illumination conditions. Visual face recognition also has difficulty in detecting disguised faces, which is critical for high-end security applications. The infrared face images are independent of ambient lighting and therefore have great advantages in poor illumination conditions, where visual face recognition systems often fail.

## 5.2 Face detection in color images

Perception of colors leads to expectations that color must be important for recognizing face images. However, the role played by color information in face recognition has been the subject of much debate. A relatively small body of research has dealt with the contribution of color in face recognition. A notable study was conducted in 1996, where it was found that observers were able to process quite normally when hue reversals of face images are done. The tasks included recognizing familiar faces or spotting differences between faces and it was concluded that color appeared to have no significant recognition advantage beyond the luminance information. In explaining these data, it was also suggested that the lack of contribution of color cues to face recognition is mainly because they do not affect shape from shading processes, which are believed to be largely color-blind. If color does play a role in face recognition, its contribution would be more evident when face landmarks or features, such as eyes or skin color are extracted for recognition. However, skin color-based segmentation has advantages over other face detection techniques since this method is almost invariant against the changes of size and orientation of face.

## 5.3 Color spaces

Color models provide a standard way to specify a particular color, by defining a 3D coordinate system and a subspace that contains all constructible colors within a particular model. Any color that can be specified using a model will correspond to a single point within the subspace it defines. Each color model is oriented towards either specific hardware (RGB,CMY,YIQ) or image processing applications (HSI). A color space is a mathematical representation of a set of colors. The three most popular color models are RGB (used in computer graphics); YIQ, YUV or YCbCr (used in video systems); and CMYK (used in color printing). However, none of these color spaces are directly related to the intuitive notions of hue, saturation and brightness. This resulted in the temporary pursuit of other models, such as HSI and HSV, to simplify programming, processing, and end-user manipulation. All of the color spaces can be derived from the RGB information supplied by devices such as cameras and scanners.

### 5.3.1 RGB model

RGB is the most common format, where the colors are represented in a cube as shown in Figure 5.1. The red, green and blue (RGB) color space is

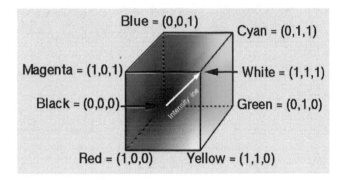

**FIGURE 5.1**: RGB color model

widely used throughout computer graphics as these are three primary additive colors (individual components are added together to form a desired color) and are represented by a three-dimensional, Cartesian coordinate system. The indicated diagonal of the cube represents an equal amount of each primary component of various gray levels. The line connecting the black point to the white point in the cube is called the intensity line.

However, RGB is not very efficient when dealing with real-world face images. All three RGB components need to be of equal bandwidth to generate any color within the RGB color cube. The result of this is a frame buffer that has the same pixel depth and display resolution for each RGB component. Also, processing an image in the RGB color space is usually not the most efficient method. For example, to modify the intensity or color of a given pixel, three RGB values must be read from the frame buffer. If the system has access to an image stored directly in the intensity and color format, some processing steps would be faster. Primarily for this reason, many video standards use luma and two color difference signals. The most common are the YUV, YIQ and YCbCr color spaces.

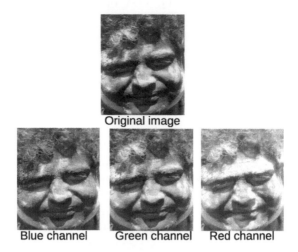

**FIGURE 5.2**: Different channels in an RGB image

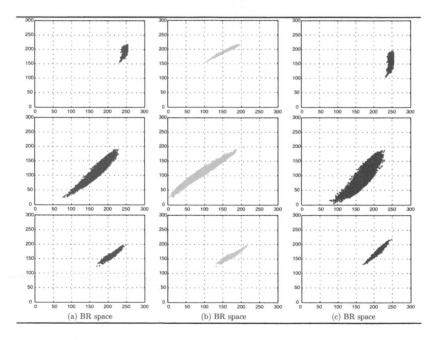

**FIGURE 5.3**: Skin color distribution of different races in different RGB color spaces. First row corresponds to Asian face, second row represents African face and third row represents Caucasian face

## 5.3.2 HSI color model

The HSI color space (hue, saturation and intensity) attempts to produce a more intuitive representation of color. The I axis represents the luminance information. The H and S axes are polar coordinates on the plane orthogonal to I. H is the angle, specified such that red is at zero, green at 120 degrees, and blue at 240 degrees. Hue thus represents what humans implicitly understand as color. S is the magnitude of the color vector projected in the plane orthogonal to I, and so represents the difference between pastel colors (low saturation) and vibrant colors (high saturation). The main drawback of this color space is that hue is undefined if saturation is zero, making error propagation in transformations from the RGB color space more complicated.

The human perception of color closely resembles the HSI color model. The I component is the average of the R,G and B components. RGB can be converted to HSI using a set of formula as

$$H \quad = \quad \theta \quad if, \quad B \leq G \qquad (5.1)$$
$$= \quad 360 - \theta \quad if, \quad B > G$$

where

$$\theta = cos^{-1}\{\frac{0.5[(R-G) + (R-B)]}{[(R-G)^2 + (R-B)(G-B)]^{0.5}}\} \qquad (5.2)$$

$$S = 1 - \frac{3}{R+G+B}[min(R,G,B)], I = 1/3(R+G+B) \qquad (5.3)$$

HSI coordinates can be converted to RGB as well.

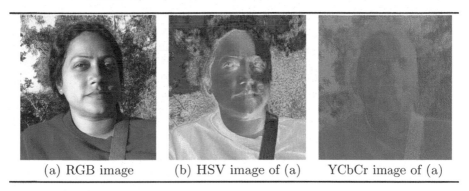

| (a) RGB image | (b) HSV image of (a) | YCbCr image of (a) |

**FIGURE 5.4**: Some example images in both RGB and HSV color space

## 5.3.3 YCbCr color space

The YCbCr color space is widely used for digital video. In this format, luminance information is stored as a single component (Y), and chrominance information is stored as two color-difference components (Cb and Cr). Cb represents the difference between the blue component and a reference value. Cr represents the difference between the red component and a reference value. Transformation from RGB to YCbCr takes an RGB input value with each component in the range [0-255] and transforms it into Y, Cb and Cr, in the ranges [0.0, 255.0], [-128.0, 127.0], and [-128.0, 127.0], respectively.

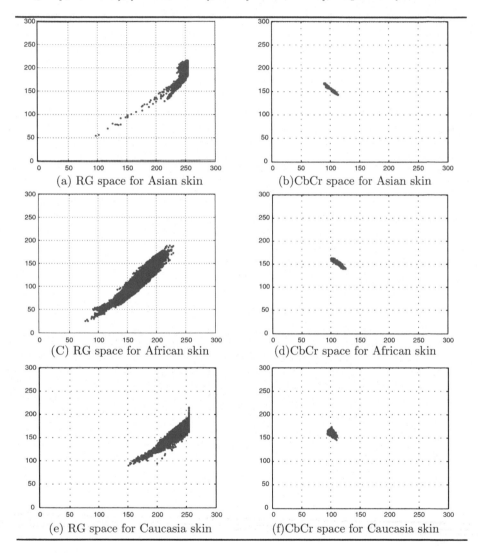

(a) RG space for Asian skin

(b)CbCr space for Asian skin

(C) RG space for African skin

(d)CbCr space for African skin

(e) RG space for Caucasia skin

(f)CbCr space for Caucasia skin

**FIGURE 5.5**: Skin color distribution in RG space and CbCr space for different races

## 5.4 Face detection from skin regions

### 5.4.1 Skin modelling

The final goal of skin color detection is to build a decision rule that will discriminate between skin and non-skin pixels. This is usually accomplished by introducing a metric, which measures distance of the pixel color to skin tone. The type of this metric is defined by the skin color modelling method.

#### 5.4.1.1 Skin color modelling explicitly from RGB space

In,[101] a skin classifier is defined explicitly through a number of rules in RGB color space as

(1) The skin classifier for skin color at uniform daylight illumination

$$R > 95 \quad AND \quad G > 40 \quad AND \quad B > 20 \quad AND$$

$$max\{R, G, B\} - min\{R, G, B\} > 15 \quad AND \quad |R - G| > 15 \quad AND$$

$$R > G \quad AND \quad R > B$$

(2) The skin color under flash-light or lateral illumination

$$R > 220 \quad AND \quad G > 210 \quad AND \quad B > 170 \quad AND$$

$$|R - G| \leq 15 \quad AND \quad R > B \quad AND \quad G > B$$

The obvious advantage of this method is simplicity of skin detection rules that leads to construction of a very rapid classifier. The main difficulty achieving high recognition rates with this method is the need to find both good color space and adequate decision rules empirically.

#### 5.4.1.2 Skin color modelling explicitly from YCbCr space

The main advantage of converting the image to the YCbCr domain is that the influence of luminosity can be removed during our image processing. In the RGB domain, each component of the picture (red, green and blue) has a different brightness. However, in the YCbCr domain all information about the brightness is given by the Y component, since the Cb (blue) and Cr (red) components are independent from the luminosity. The Cb and Cr components give a good indication on whether a pixel is part of the skin or not. Figure 5.6 indicates strong correlation between Cb and Cr values for skin image (in blue). Figure 5.6 also indicates the distribution of non-skin image (in red)in Cb-Cr space. Hence by applying maximum and minimum threshold values for both Cb and Cr components skin part can be easily segmented from background image. Figure 5.7 shows the skin segmentation by applying the threshold of Cb,Cr components.

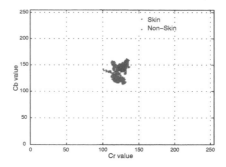

**FIGURE 5.6**: Skin and non-skin part distribution in Cb-Cr space.

**FIGURE 5.7**: Skin part detection in Cb-Cr space.

## 5.5 Probabilistic skin detection

Probabilistic skin detection is based on Bayes rule. In this method of approach Bayes rule is applied for each pixel and the probability of being a skin part is judged. Probability of being skin for a given RGB value can be expressed by conditional probability as $P(skin|RGB)$. This $P(skin|CS)$[1] can be obtained from a given scene by applying Bayes rule as

$$P(skin|CS) = \frac{P(CS|skin)P(skin)}{P(CS)} \tag{5.4}$$

Specifically, for RGB color space, Bayes rule becomes

$$P(skin|RGB) = \frac{P(RGB|skin)P(skin)}{P(RGB)} \tag{5.5}$$

Now $P(RGB)$ can be written as

$$P(RGB) = P(RGB|skin) * P(skin) + P(RGB|\overline{skin}) * P(\overline{skin}) \tag{5.6}$$

where $P(\overline{skin})$ represents probability of non-skin. For simplicity $P(skin)$ and $P(\overline{skin})$ can be taken as 0.5 each. $P(RGB|skin)$ can be formulated as

$$P(RGB|skin) = P(R|skin) * P(G|skin) * P(B|skin) \tag{5.7}$$

---

[1] CS stands for color space. CS may be RGB or HSV or YCbCr or any other color format.

where the color channels are considered to be independent to each other. $P(RGB|skin)$ is the probability that we will observe RGB, when we know that the pixel is a skin pixel. Figure 5.8 illustrates how to select the skin and non-skin sub-windows from an original scene for training of $P(RGB|skin)$ and $P(RGB|\overline{skin})$.

**Original scene**

**Skin portion**          **Non-skin portion**

**FIGURE 5.8**: Subwindow of skin and nonskin part for an example image

With the subwindows a simple Gaussian model can be estimated for class conditional pdfs $P(RGB|skin)$ and $P(RGB|\overline{skin})$. This is a parametric method. The parameters used to generate the Gaussian models are mean and standard deviation. In the case of the unimodal Gaussian model, the skin class conditional pdfs have the following forms as

$$p(RGB|skin) = (2\pi)^{d/2}|C_{skin}|^{-0.5}exp\{-\frac{1}{2}(x - m_s)^T C_{skin}^{-1}(x - m_s)\} \quad (5.8)$$

where $d$ is the dimension of the feature vector, $m_s$ is the mean and $C_{skin}$ is the covariance matrix of the skin class. Similarly for the non-skin class we have

$$p(RGB|\overline{skin}) = (2\pi)^{d/2}|C_{\overline{skin}}|^{-0.5}exp\{-\frac{1}{2}(x - m_{ns})^T C_{\overline{skin}}^{-1}(x - m_{ns})\} \quad (5.9)$$

where $m_{ns}$ represents the mean of the non-skin class features. Assuming the colors are independent, the class conditional pdf $P(RGB|skin)$ can be rewritten for a different color channel as

$$P(RGB|skin) = P(Red|skin) * P(Green|skin) * P(Blue|skin) \quad (5.10)$$

or

$$P(RGB|skin) = N(Red, m_r, \sigma_r) * N(Green, m_g, \sigma_g) * N(Blue, m_b, \sigma_b)$$

$$= \frac{1}{\sqrt{2\pi}\sigma} exp\{-\frac{1}{2\sigma^2}(R - m_r)^2\} \times$$

$$\frac{1}{\sqrt{2\pi}\sigma} exp\{-\frac{1}{2\sigma^2}(G - m_g)^2\} \times$$

$$\frac{1}{\sqrt{2\pi}\sigma} exp\{-\frac{1}{2\sigma^2}(B - m_b)^2\} \tag{5.11}$$

Estimating both $P(RGB|skin)$ and $P(RGB|\overline{skin})$ from the training data, for a given test data, Bayes classifier is used to evaluate $P(skin|RGB)$. Now apply the skin model for every pixel of the given image and evaluate

$$\frac{P(RGB|skin)}{P(RGB|\overline{skin})} > \tau \tag{5.12}$$

to detect the skin portion in the image. $\tau$ is the threshold. Figure 5.9 shows results obtained by the probabilistic skin detection method.

**FIGURE 5.9**: Skin color detection using probabilistic skin model application

## 5.6 Face detection by localizing facial features

It has been shown in the previous sections how to segment human skin from an unconstrained background by applying different skin color models and probabilistic skin modelling. Having obtained the skin part, face detection can be done in several ways. One of the ways of face detection [102] will be described in this section. By this method the important facial features are localized and then the face part is extracted.

Among the various facial features, eyes and mouth are the most prominent features for recognition and estimation of the 3D head pose. In this section the eye and mouth are localized from the detected skin part. To detect eye and mouth the information of both luma and chroma components is exploited along with morphological analysis.

### 5.6.1 EyeMap

In [102] two separate eye maps are developed using (1) a chrominance component and (2) a luminance component. The eye map from the chroma is based on the observation that high Cb and low Cr values are found around the eyes. It is constructed by

$$EyeMapC = \frac{1}{3}\{C_b^2 + \tilde{C}_r^2 + \frac{C_b}{C_r}\} \tag{5.13}$$

where $C_b^2, \tilde{C}_r^2, \frac{C_b}{C_r}$ all are normalized to the range $[0, 255]$ and $\tilde{C}_r$ is the negative of $C_r$, i.e., $\tilde{C}_r = 255 - C_r$. Detailed construction of EyeMapC is shown in Figure 5.10. Since the eyes usually contain both dark and bright pixels in the luma

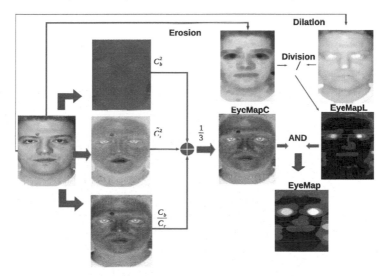

**FIGURE 5.10**: Construction of EyeMap

component, gray-scale morphological operators (for example, can be designed to emphasize brighter and darker pixels in the luma component around the eye regions. EyeMapL is developed by gray-scale dilation and erosion using the following equation,

$$EyeMapL = \frac{Y(i,j) \oplus s(i,j)}{Y(i,j) \ominus s(i,j) + 1} \tag{5.14}$$

where $\oplus$ is gray-scale dilation and $\ominus$ is gray-scale erosion operations on function $Y(i,j) : F \subset \Re^2$. The luma component in YCbCr color space, using the structuring element $s(i,j) : G \subset \Re^2$, are defined as follows:

$$Y \oplus s(i,j) = \max\{Y(i-c,j-r)+s(c,r)\}; (i-c,j-r) \in F, (c,r) \in G \tag{5.15}$$

$$Y \ominus s(i,j) = \min\{Y(i-c,j-r)+s(c,r)\}; (i-c,j-r) \in F, (c,r) \in G \tag{5.16}$$

$$s(i,j) = \left\{ \begin{array}{ll} |\sigma| \cdot \left( |1 - (R(i,j)/s)^2|^{-1/2} - 1 \right); & R \leq |\sigma| \\ -\infty; & R > |\sigma| \end{array} \right.$$

$$R(i,j) = \sqrt{i^2 + j^2}$$

Detailed construction of EyeMapL from the luma is illustrated in Figure 5.10. The eye map from the luma is combined with the eye map from the chrominance component by an AND operation to yield the EyeMap in the target image.

$$EyeMap = \{EyeMapC\} \quad \textbf{AND} \quad \{EyeMapL\} \tag{5.17}$$

The resulting eye map is then enhanced by dilation and erosion operation and the other facial landmarks are suppressed. The resulting eye map is illustrated in Figure 5.10.

## 5.6.2   MouthMap

The color of the mouth region contains a stronger red component and weaker blue component than other facial regions. Hence, the chrominance

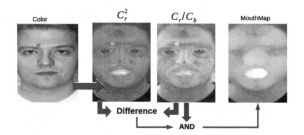

**FIGURE 5.11**: Construction of MouthMap

component Cr is greater than Cb in the mouth region. The mouth has a relatively low response in the $C_r/C_b$ feature, but it has a high response in $C_r^2$. The mouth map can be constructed as

$$MouthMap = C_r^2 \cdot (C_r^2 - \eta C_r/C_b)^2 \tag{5.18}$$

where

$$\eta = 0.95 \frac{1/n \sum\limits_{(i,j \in FG)} C_r^2(i,j)}{1/n \sum\limits_{(i,j \in FG)} \frac{C_r(i,j)}{C_b(i,j)}} \tag{5.19}$$

where $C_r^2, C_r/C_b$ are normalized to the range $[0, 255]$. $\eta$ estimates the ratio of the average of $C_r^2$ to the average of $C_r/C_b$. Figure 5.11 illustrates the steps involved in finding the mouth in a target image. Having obtained an eye map and mouth map, the position of both the eyes and mouth can be estimated by computing the highest pixel position in the target image. From the pixel's position iteratively these facial landmarks are identified. These landmarks indicate the face boundary roughly as shown in Figure 5.12

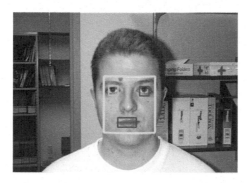

**FIGURE 5.12**: Example of face detection in a color image. First, the facial landmarks are found and then the face part is extracted (shown with a big square box)

---

MATLAB Code for face detection in color image

---

```
%% Face detection in color image

im = imread("....jpg");
im = imresize(im,[340,480]);
img = rgb2gray(im);
ycbcr = rgb2ycbcr(im);
Cb = ycbcr(:,:,2);
Cr = ycbcr(:,:,3);
for ic = 1:size(im,1)
    for ik = 1:size(im,2)

        if (Cr(ic,ik)>135 && Cr(ic,ik)<165 && Cb(ic,ik)>...
                110 && Cb(ic,ik)<130)
            img(ic,ik) = 255;
        else
            img(ic,ik)=0;
        end
    end
end
clear Cb Cr ycbcr
st = strel("square",15);
img = imerode(img,st);
st = strel("square",3);
img = imdilate(img,st);
imbw = im2bw(img);
```

```
[L,n] = bwlabel(imbw,4);
tmp =0;
for ic = 1:n
    cc(ic) = size(find(L==ic),1);
    if tmp < cc(ic)
        indx =ic;
        tmp = cc(ic);
    end
end
figure
imshow(img);
[h1,h2] = find(L==indx);
imCrop = im(min(h1):max(h1),min(h2):max(h2),:);
figure
imshow(imCrop);
ycbcr  = rgb2ycbcr(imCrop);
Cb = im2double(ycbcr(:,:,2));
Cb2 = Cb.^2;
Cr = im2double(ycbcr(:,:,3));
figure
imshow(Cb./Cr,[]); title("cbByCr");
Cr2 = Cr.^2;
nCr2 = (1-Cr).^2;
figure, imshow(nCr2,[]); title("nCr2");
EyeMapC = (1/3)*(Cb2 + (1-Cr).^2 + (Cb./Cr));
figure; imshow(EyeMapC,[]);title("EyeMapC");
Y = ycbcr(:,:,1);
figure,imshow(Y,[]); title("Y");
Y = Y.*(255/max(max(Y)));
s = strel("ball",5,5);
Yd = imdilate(Y,s);
figure,imshow(Yd,[]);title ("dilate")
Ye = imerode(Y,s);
figure,imshow(Ye,[]);title ("erode")
EyeMapL = Yd./(Ye+1);
figure; imshow(EyeMapL,[]);title("EyeMapL");
EyeMapL = im2double(EyeMapL);
EyeMapL = EyeMapL.*(255./(max(max(EyeMapL))));
EyeMapC = EyeMapC.*(255./(max(max(EyeMapC))));
AndEye = EyeMapL.*EyeMapC;
s = strel("ball",5,5);
Eye = imerode(AndEye,s);
s = strel("ball",11,11);
Eye = imdilate(Eye,s);
figure; imshow(Eye,[]);title("Eye");
figure; imshow(Cb2);title("Cb2");
figure
Cr2 = Cr2.*(255./(max(max(Cr2))));
imshow(Cr2,[]);title("Cr2");
```

```
CrByCb = Cr./Cb;
CrByCb = CrByCb.*(255./(max(max(CrByCb))));
figure; imshow(CrByCb,[]);title("CrByCb");
eta = 0.95 * (sum(sum(Cr.^2))/(sum(sum(Cr/Cb))));
MouthMap =Cr2.*(abs(Cr2-eta*CrByCb));
figure, imshow(abs(Cr2-eta*CrByCb),[]),title("Diff");
s = strel("ball",5,5);
Mouth = imerode(MouthMap,s);
s = strel("ball",11,11);
Mouth = imdilate(Mouth,s);
figure; imshow(Mouth,[]);title("Mouth");
[h7,h8] = find(Mouth==max(max(Mouth)));
mr = h7+min(h1);
mc = h8+min(h2);
figure; imshow(im);hold on
rectangle("Position",[mc-30,mr-10,40,20],...
    "LineWidth",3,"EdgeColor","b")
[h3,h4]=find(Eye==max(max(Eye)));
rer = h3+min(h1);
rec = h4+min(h2);
rectangle("Position",[rec-10,rer-10,20,20],...
    "LineWidth",3,"EdgeColor","r")
Eye(h3-20:h3+20,h4-20:h4+20)=0;
[h5,h6]=find(Eye==max(max(Eye)));
ler = h5+min(h1);
lec - h6+min(h2);
rectangle("Position",[lec-10,ler-10,20,20],...
    "LineWidth",3,"EdgeColor","g")
if lec<rec
    rectangle("Position",[lec-20,ler-30,110,120],...
        "LineWidth",4,"EdgeColor","c")
else
    rectangle("Position",[rec-20,rer-30,110,120],...
        "LineWidth",4,"EdgeColor","c")
end
```

## 5.7   Face detection in infrared images

Face recognition in the infrared domain has received relatively little attention in comparison to recognition done with face images taken in the visible spectrum. Despite the success of automatic face recognition techniques in many practical applications, recognition based only on the visual spectrum has difficulties performing consistently under uncontrolled operating environments as the performance is sensitive to variations in illumination conditions. Moreover, the performance degrades significantly when the lighting is dim or when it is not uniformly illuminating the face. Even when a face is well lit, shadows, glint, makeup and disguises can cause errors in locating the feature points in face images.

The infrared spectrum of an electromagnetic wave is divided into four bandwidths: near-IR (NIR), short-wave-IR (SWIR), medium-wave-IR (MWIR) and long-wave IR (thermal IR). Face images at long IR represent the heat patterns emitted from the face and thus are relatively independent of ambient illumination. Infrared face images are unique and can be regarded as thermal signature of a human. Because of these reasons, infrared face recognition is useful under all lighting conditions including total darkness and also when the subject is wearing a disguise, and therefore is of particular interest in high-end security applications. Symptoms such as alertness and anxiety reflected in the face can easily be detected as redistribution of blood flow in blood vessels, causing abrupt changes in the local skin temperature.

One of the first infrared face recognition systems was introduced in 1997. It focused on the lighting problem in face recognition and suggested infrared face recognition as a solution. An important property of thermal face images, that IR images are affected by changes of pose or facial expression, was investigated. Comparison of thermal, visible and range images can be done using the amount of variation as a comparison criterion.

Appearance-based approaches are commonly used for IR face recognition systems. In contrast to visual face recognition algorithms that mostly rely on the eye location, thermal IR face recognition techniques present difficulties in locating the eyes. Initial research approach to thermal face recognition extracts and matches thermal contours for identification. Such techniques include elemental shape matching and the eigenface method. Variations in defining the thermal slices from one image to another has the effect of shrinking or enlarging the resulting shapes, while keeping the centroid location and other features of the shapes constant. Perimeter, area, $x$ and $y$ coordinates of the centroid, minimum and maximum chord length through the centroid and between perimeter points and standard deviation of that length are considered as features. Such an automated face recognition system using elemental shapes in real time has reported high accuracy for cooperative access control applications. A non-cooperative, non-real-time, faces-in-the-crowd version of thermal face recognition also achieved very high accuracy with almost no false positives when trained with a medium-sized face database.

## 5.8 Multivariate histogram-based image segmentation

Feature extraction is governed by several segmentation algorithms for the color images namely clustering methods, histogram based region growing methods [103], etc. Some works are reported [104] on finding peaks and valleys in bivariate and multivariate histograms. The early work in this area was related to the development of an algorithm involving parent-child relationship between bins in a bivariate histogram [105]. The number of clusters in the method depends on the choice of control parameters. Subsequently, few other works were reported later on[106].

The basic philosophy behind the methods is guided by the observations from a single class/region which tends to form a cluster in the feature space i.e., a peak in the multi-dimensional histogram . Then the analysis is conducted to identify suitable boundaries of these peaks. However, implicit in the peak search is the knowledge of the number of segments and therefore, it is necessary to search the number of peaks in the histogram. Many peak detection method are available [105]and [107], however, some pre and post processing of the data needs to be incorporated to obtain the clusters. These clusters are then used for facial feature extraction, dimensionality reduction and classification [53],[78],[108].

A bivariate histogram provides a histogram corresponding to two variables and hence the bivariate histogram is a matrix. If we represent the matrix as $A$ and the *(i,j)* th element of the matrix as *a(i,j)*, then *a(i,j)* denotes the number of pixels in the image having the gray-value $i$ for the first variable and the gray value $j$ for the second variable. The input for finding a multivariate histogram consists of images for three color channels R, G and B and four channels R, G, B and IR for thermal imagery corresponding to the same person and taken at the same time and of the same size $M \times N$. In case of thermal imagery, $4_{C_2}$ bivariate histograms, corresponding to the band pairs (R,G), (R,B), (R,IR), (G,B), (G,IR) and (B,IR) are available.

A trivariate histogram index Trivariate histogramis represented as $H$ and each element of it is represented as $h(i,j,k)$, where $h(i,j,k)$ denotes the number of pixels in the image having gray-value $i$ for the first variable, gray value $j$ for the second variable and gray-value $k$ for the third variable. For thermal imagery, *4*-trivariate histograms can be constructed, corresponding to the band triplets (R,G,B), (R,G,IR), (R,B,IR) and (G,B,IR). Any multivariate histogram can be defined similarly by a single *4*-variate histogram including a thermal image. On the other hand, there are *4*-univariate histograms corresponding to the four variables for thermal imagery.

Significance of peaks and valleys of histograms are related in the formation of segments. Similar to the analysis of univariate histograms, number of peaks or the modes in a multivariate histogram signifies the number of clusters. The formed clusters in a multivariate histogram are basically the color clusters. However, there may also be spurious peaks, which needs to be eliminated. Valleys also play an important role in histogram thresholding and decides the cluster boundary. For one dimensional histogram, the valley point separates two modes. For a bivariate histogram, a valley is a line (or curve) separating the cluster regions. In case of a trivariate histogram, a valley is a plane separating two clusters.

### 5.8.1    Method for finding major clusters from a multivariate histogram

For the sake of convenience, the method of segmentation for a trivariate histogram is described. This can, however, be easily extended to any number of variables. The complete method of segmentation for an input image is presented using the following algorithmic steps. The steps are (a) smoothing of the multivariate histogram , followed by (b) finding the peaks and valleys and then (c) detection of the major clusters in the histogram.

**Input** : Let the given image be represented by $I$ and the color vector of the $(i, j)$th pixel be represented by $I(i, j)$. The corresponding three variables are R, G and B. Let also $min_l$ and $max_l$ denote the minimum and maximum gray values of $l$, where $l$ = R or G or B. Let $R(i, j), G(i, j)$ and $B(i, j)$ denote the color intensity values of $(i, j)$th pixel for the colors R, G and B respectively. Thus, $I(i, j) = (R(i, j), G(i, j), B(i, j))$. The corresponding histogram be denoted by $H$, and $h(p, q, r)$ denote the number of pixels having the R value as $p$, G value as $q$ and B value as $r$. Note that $min_R \leq p \leq max_R$, $min_G \leq q \leq max_G$, and $min_B \leq r \leq max_B$.

**Step 1**: Multivariate histogram smoothing

A smoothing methodology for the removal of local variations in histogram is used. Let, after smoothing, the new smoothed histogram be represented by $H_1$, and $h_1(p, q, r)$ denotes the value of $(p, q, r)$ in $H_1$. Then,

$$h_1(p, q, r) = \frac{1}{27} \sum_{k=r-1}^{r+1} \sum_{j=q-1}^{q+1} \sum_{i=p-1}^{p+1} h(i, j, k) \qquad (5.20)$$

For every $(i, j, k)$ of $H$, the above operation needs to be performed. If the maximum and minimum values are the $f_{max}$ and the $f_{min}$ then for each dimension, the smoothing operation needs to be performed from $f_{min}$ to $f_{max}$. Note that, when $p = min_R$ or $max_R$, $h_1(p, q, r) = h(p, q, r)$. Same convention will hold when $q = min_G$ or $max_G$ or $r = min_B$ or $max_B$.

**Step 2**: Finding peaks and valleys of the histogram

Once the smoothing operation is accomplished, the smoothed multivariate histogram is used to find peaks and valleys. The process generates the tree structure by examining neighbors of each bin and then the largest bin, i.e. the bin with the largest number of elements is selected. Then the links are established. If the current bin has the same value as the largest neighbor, one of them is selected as the father and the link is established. If the current bin is the largest among its neighbors, then the search is stopped for the current bin. Each histogram bin is connected to a bin that has the largest value in its neighborhood. Each bin is connected to a single parent bin by a path.

**Step 3**: Detection of major clusters in the histogram

A post processing step is developed to detect the major clusters in the histogram. The number of bins in every cluster is counted. The two clusters having the two largest number of bins are considered. If the number of determined peaks in the histogram is equal to or less than the desired number of segments, $K$, the algorithm is terminated. Otherwise some of the local peaks are eliminated iteratively until their number reduces to $K$. To facilitate this process, each peak is attributed with the value of the sum of its children while the child bins are set to zero. After performing the post processing, the segmented image is created.

## 5.8.2 Experiments and results on the color and IR face image datasets

For the application of the developed technique for the detection of facial features, the cluster with the largest number of bins corresponds to the skin color. The cluster of interest is the second largest cluster containing the feature sets (two eyes, nose and mouth) as the connected components. The cluster with the largest number of members (skin portion of face) is thus removed and the second largest cluster, which normally contains the portions corresponding to two eyes, nose and mouth is considered as the three basic features of the face.

Two stages of experiments are carried out to establish the process of multivariate image segmentation. In stage 1, the proposed algorithm is applied on the histogram(s) to obtain the facial features as segmented parts of the image. In stage 2, the dimensionality reduction and the classification are performed on the segmented parts using the nearest neighbor classifier. That is, in the stage 2, the utility of the proposed method for segmentation is verified for classification accuracy. The details of these stages are discussed below.

At Stage 1 of the experiment, the vector values (R,G,B) for each pixel of AR imagery or the vector values (R,G,B,IR) of IRIS thermal/visual dataset, for each face image are provided as input for multivariate histogram segmentation algorithm. The features of the face images are segmented and initially three largest segments are formed. The second largest segment is normally selected to contain the feature set.

In Figure 5.13, the images in the first column are representative RGB images from AR data. The images in the second column are the images obtained containing the extracted features. In Figure 5.14, the images in the first column are representative color images from visual/IRIS data set. The images in the second column are the IR counterparts of the images of the first column. The images in the third column depict the feature extracted images obtained due to the segmentation procedure undertaken.

The following observations are made:

1. For AR face dataset, for each image, 3 bivariate histograms can be formed. As the separation between blue and red wave lengths is high, the experiment justifies that the bivariate histogram with red and blue variables provides better results in distinguishing the facial features more effectively.

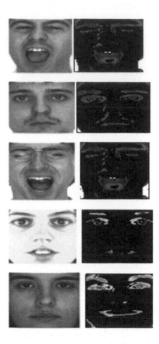

**FIGURE 5.13**: RGB face images from AR dataset and corresponding extracted feature segments using bivariate histogram

**FIGURE 5.14**: RGB and IR face images from IRIS visual/thermal dataset and corresponding extracted feature segments using trivariate histogram

2. For the thermal visual dataset, among the six bivariate histograms, the histogram corresponding to the combination of IR and blue bands is found to provide better results,because of same reason as stated. For 3D histograms the best combination is found to be the histogram with R, G and IR bands.

3. Inclusion of IR band is found to increase the quality of the feature set. That is, the results of the experiment on 3D histogram containing IR band along with R and B bands is found to be better than other 2D and 3D combinations of histograms.

4. In case of AR data set, it may be noted that the images with different illumination is included in the tests.

5. It may also be stated that the feature in the IRIS-IR data set is illumination invariant. However, the folder '2on' contains images with different illumination.

During stage 2 of the experiment, the utility of the segmentation for the recognition of faces is established. The features like eyes, nostrils and lip portions are extracted to obtain T structure and the T structure is dimensionally reduced by subspace methods [109] [99]. The recognition rates using these reduced features are found to be better than the dimensionality reduction schemes *without* using the T structure. The feature sets are automatically segmented out from the other face portions in Stage I and those are used for dimensionality reduction using subspace based methods (PCA and 2DPCA). The nearest neighbour algorithm is used for classification [73].

### 5.8.3 Utility of facial features

Two experiments are carried out to establish the utility of the method. In the first experiment, only the obtained facial features are used for dimensionality reduction and classification. In the second experiment, the complete images are used for dimensionality reduction and classification. The training and test sets are selected from IRIS and AR dataset. After the training-test division of datasets, both the datasets are processed for segmenting out the face features.

PCA and 2DPCA are applied on the training set of the original and the segmented image set. The number of dimensions after reduction using PCA (or 2DPCA) for IRIS dataset is 180, when the segmented images are only used or the full image are used for reduction. The number of dimensions after reduction for AR dataset is 160. The reduced values of dimension are intentionally considered to be the same to maintain parity during comparison. The test image set of original and the segmented test faces are projected to the corresponding face spaces. The nearest neighbour classifier is used for classification.

Table 5.1 shows the results on both the original images and the images generated by the algorithm. The experimental results in Table 5.1 show that the selected feature set images outperform the original image set when the PCA based face recognition method is applied.

**TABLE 5.1**: Comparative study of recognition rates using full image and segmented images

| Method applied | Dataset | Recognition rate using segmented image |
| --- | --- | --- |
| PCA | IRIS | 91 |
| 2DPCA | IRIS | 94.5 |
| PCA | AR | 94.8 |
| 2DPCA | AR | 95 |

The basic observation on the technique of the multivariate histogram segmentation is that (R,B) offers the best performance among all 2-feature combinations on AR dataset, (IR,B) works the best among all 2-feature combinations on IRIS dataset, and (IR,G,B) performs best among all 3-feature combinations for IRIS dataset. In these cases the feature set is more distinctly found.

There are 6 pairs of color bands for IRIS data and the correlation coefficient can be calculated for each pair. Among these pairs, it is possible to select those bands which preserve the maximum information. Such combinations are more suitable for constructing the multivariate histograms than other correlated bands. Since B and IR are maximally separated according to their wavelength than any other combination, B and IR combination is chosen for IRIS data. Similarly, R and B are chosen for the AR data.

It may be noted that, in each data set, images with different illuminations are included. The obtained results using the technique are found to provide better recognition rates even under variations in illumination. Thus the proposed is seen to overcome the restriction of illumination variations to a great extent.

The approach for segmenting the image using histograms is unsupervised and is very close to the dominant colors present in the original image. Each one of the four univariate histograms (corresponding to the four channels) has a single mode and thus segmentation based upon finding valleys in a histogram becomes a difficult task. Further, a smoothing technique is applied on histograms to remove the spurious peaks. The peaks are used to get the major segments and one can use the valley regions for edge detection purpose. The method of facial feature extraction using histograms is independent of feature position. It is also illumination invariant to a great extent. The performance of the 2D histogram processing for the face images is superior to the 1D histogram because more information is used, and hence the valley regions are much clearer. The performance of the method is verified on an artificial dataset and applied on the color AR face and IR face datasets. The segmented face images thus obtained, contain the skin portion.

In 3D case, when combinations of RGB color channels are taken, the clusters are distinguishable. Moreover, IR channel gives better and distinct results as thermal imageries are illumination invariant. IR images in combination with other color bands used in the multivariate histogram segmentation results in considerable improvement in generating feature set. However, if IR face images are with spectacles on eye, then the univariate segmentation algorithms fails to detect eye portions.

# Chapter 6

## Intelligent face detection

## 6.1 Introduction

Intelligent approaches for face detection and recognition utilize tools of artificial neural networks(ANNs) and machine learning techniques to detect and recognize faces. The development of an intelligent face recognition system requires providing sufficient information and meaningful data during machine learning of a face. However, the application of neural networks for face detection tasks is difficult and more challenging than face recognition, because of the difficulty in characterizing prototypical nonface images. Unlike face recognition, in which the classes to be discriminated are different faces, the two classes to be discriminated in face detection are images containing faces and images not containing faces. It is hard to get a representative sample of non-face images. In [110] the problem of using a huge training set for nonfaces is avoided by selectively adding images to the training set as training progresses. This bootstrap method reduces the size of the training set needed. The use of arbitration between multiple networks and heuristics to clean up the results significantly improves the accuracy of detection. A view-based approach to face detection [110] uses an artificial neural network to represent each view. Before going into the details of face detection using the neural network this chapter initially details the multilayer perceptron model of the artificial neural network (ANN) and the backpropagation algorithm associated with neural network training.

---

## 6.2  Multilayer perceptron model

An artificial neural network (ANN) consists of an ensemble of highly parallel and interconnected processing elements (PEs) called neurons, analogous to the biological neural network. The multilayer perceptron model is a basic approach of modelling and analysis of such an interconnected system. The connection between a pair of PEs, i.e., $PE_i$ and $PE_j$, has an associated strength or synaptic weight of an adaptive coefficient denoted by $w_{ij}$. Positive weights represent excitatory connection, which increases the strength of the connection, while negative weights represent inhibitory connections which decreases its strength. The value $w_{ij} = 0$ means that there is no interconnection between $PE_i$ and $PE_j$.

A PE has several input connections and a single output connection that again branches out to several collateral connections, each carrying the same signal. A single PE can be represented as an adder of several inputs multiplied by their synaptic weights, followed by a transfer function which is usually described by a non-linear relationship to determine the threshold output.

Some of the features of ANN are:

1. Adaptability: The property of ANN to automatically learn to respond to a newly encountered input pattern by adapting the synaptic weights connecting the PEs in the network is termed as adaptability.

2. Distributed Storage and Associated Memory: Any information within an ANN is distributed and encoded in the connections and not stored in a specific location. Also, ANN has an associative memory which accesses information by content.

3. Fault Tolerance: Since information is distributed across many PEs in the network, information is not lost due to damage of a few PEs or links. ANNs have the ability to tolerate hardware malfunctions and are hence, quite fault-tolerant.

Training the ANN can be either through the supervisory or unsupervised mode. Based on this classification, ANN networks like the perceptron model, multilayer perceptron, Hopfield model and Boltzmann machine belong to the class of supervised ANNs, whereas Kohonen's self-organizing feature map and the Carpenter/Grossberg model belong to the class of unsupervised ANNs. Supervised ANNs require supervised training wherein the ANN is supplied with a known sequence of inputs $(x_1, x_2, ...x_k....)$ and the desirable or correct outputs $(y_1, y_2, ...y_k....)$ are expected after processing in the ANN. For this, the ANN undergoes an iterative process such that the output obtained is compared with the desired output and the difference, if any, is corrected by using some learning algorithm by modifying the synaptic weights. The process is repeated till the actual output reaches an acceptable value close to the desired output.

The perceptron model is a single-layer network consisting of one or more PEs. It is a supervised ANN model, in which the input, either binary or continuous-value, are used to train the network. It is basically a form of linear classifier, which maps its input $x$ to the output $f(x)$. When the input pattern $(x_0, x_1, ...x_{n-1})$ is applied at the input, it gets multiplied with the corresponding interconnecting weights $(w_0, w_1, ...w_{n-1})$. The addition of all the weighted inputs is thresholded to represent output $y$ of the PE. So if $\theta$ is the threshold function for the PE, then the output $y$ may be written as

$$y(t) = \Sigma_{i=0}^{n-1} w_i(t)x_i(t) - \theta \qquad (6.1)$$

The output $y$ of the ANN is +1, if the weighted sum is greater than the threshold value and the input belongs to Class 1, whereas the output $y$ of the ANN is -1, if the weighted sum is less than the threshold value and the input then belongs to Class 0. The output $y$ of the ANN at time $t$ as written in Equation 6.1 becomes

$$y(t) = +1 ... if, y(t) > 0$$
$$= -1 ... otherwise \qquad (6.2)$$

The synaptic weights at time $t + 1$ are modified on the basis of the difference $(\triangle)$ between the desired $(d_i(t))$ and obtained output $(y_i(t))$ in such a way that

$$w_i(t + 1) = w_i(t) + \eta\triangle \qquad (6.3)$$

where $0.1 < \eta < 1.0$ controls the adaptation rate of the weights.

For a one-layer neural network of $N$ neurons or PEs, there would approximately be $N^2$ interconnections. Hence the state of the $i$-th neuron can be expressed as

$$y_j(t) = \Sigma_{j=0}^{n-1} w_{ij}(t)x_i(t) - \theta_j \qquad (6.4)$$

where $W_{ij}$ is the interconnection weight matrix $W_{ij}$ (IWM) between the $i$-th and $j$-th neuron. The first term of Equation 6.4 is basically a matrix-vector outer-product.

The multilayer perceptron is a feed forward ANN, wherein the first layer is the input layer whereas the $n$-th layer is the output layer. All layers in between the input and output layers are hidden layers. A node in a particular layer is connected to all nodes of the previous layer as well as all nodes of the subsequent layer. The input-output mapping of the multilayer perceptron is shown in Figure 6.1. The number of hidden layers and the number of nodes in the hidden layers are determined by the proper internal representation of the input patterns best suited for classification of linearly inseparable problems.

For the perceptrons in the input layer, the linear transfer function is used. For the perceptrons in the hidden layer and the output layer sigmoidal functions are used.

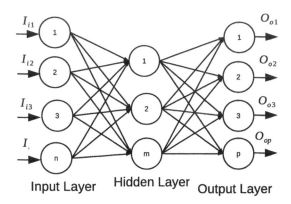

**FIGURE 6.1**: Multilayer perceptron

A three-layer network is shown in Figure 6.1. The activity of the neurons in the hidden layer is determined by the activities of the neurons in the input layer and the connecting weights between input and hidden units. Similarly, the activity of the output units depends on the activity of neurons in the hidden layer and the weight between hidden and output layers. Neurons in the hidden layer are free to construct their own representations of the input.

## 6.2.1   Learning algorithm

Consider the network shown in Figure 6.2 where the subscripts $i, h, o$ denote input, hidden and output neurons.

- Step1: Normalize the inputs and outputs with respect to their maximum values. It is proved that the neural networks work better if inputs and outputs lie between 0 to 1. For each training pair assume there are $p$ inputs given by $I_p$ and $n$ outputs $O_n$ in normalized forms.

- Assume the number of neurons in the hidden layer to lie between $1 < m < 2p$.

- $V$ represents the weights of synapses connecting input neurons and hidden neurons and $W$ represents the weights of synapses connecting hidden neurons and output neurons. Initialize the weights to small random values usually from -1 to +1.

- For the training data, present one set of inputs and outputs. Present the pattern to the input layer $I_i$ as inputs to the input layer. By using the

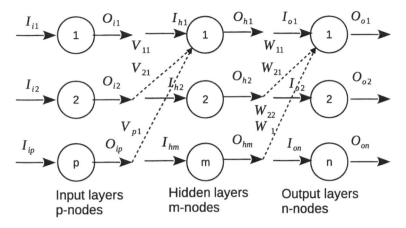

Input layers
p-nodes

Hidden layers
m-nodes

Output layers
n-nodes

**FIGURE 6.2**: Multilayer feedforward backpropagation network

linear activation function, the output of the input layer may be evaluated as

$$O_{i_{p \times 1}} = I_{i_{p \times 1}} \tag{6.5}$$

- Compute the inputs to the hidden layer by multiplying corresponding weights of synapses as

$$I_{h_{m \times 1}} = V_{m \times p}^{T} O_{i_{p \times 1}} \tag{6.6}$$

- Let the hidden layer units evaluate the output using the sigmoidal function as

$$O_h = \begin{pmatrix} \cdot \\ \cdot \\ \frac{1}{(1+e^{-I_{hk}})} \\ \cdot \\ \cdot \end{pmatrix} \tag{6.7}$$

- Compute the inputs to the output layer by multiplying corresponding weights of synapses as

$$I_{o_{n \times 1}} = W_{n \times m}^{T} O_{h_{m \times 1}} \tag{6.8}$$

- Let the output layer units evaluate the output using the sigmoidal function as

$$O_o = \begin{pmatrix} \cdot \\ \cdot \\ \frac{1}{(1+e^{-I_{oj}})} \\ \cdot \\ \cdot \end{pmatrix} \tag{6.9}$$

This equation represents the network output.

- Calculate the error and the difference between the network output and the desired output as for the $k$th training set as

$$e = \frac{\sqrt{\sum(T_j - O_{oj})^2)}}{n} \qquad (6.10)$$

Find $d$ as

$$d = \begin{pmatrix} \cdot \\ \cdot \\ (T_k - O_{ok})O_{ok}(1 - O_{ok}) \\ \cdot \\ \cdot \end{pmatrix} \qquad (6.11)$$

- Find $Y$ matrix as

$$Y = O_h d \qquad (6.12)$$

- Find

$$[\Delta W]^{t+1} = \alpha[\Delta W]^t + \eta Y \qquad (6.13)$$

- Find

$$e = W d \qquad (6.14)$$

$$d^* = \begin{pmatrix} \cdot \\ \cdot \\ e_k(O_{hk}(1 - O_{hk}) \\ \cdot \\ \cdot \end{pmatrix} \qquad (6.15)$$

Find $X$ matrix as

$$X = O_i d^* = I_i d^* \qquad (6.16)$$

- Find

$$[\Delta V]^{t+1} = \alpha[\Delta V]^t + \eta X \qquad (6.17)$$

- Find

$$[V]^{t+1} = [V]^t + [\Delta V]^{t+1} \qquad (6.18)$$

$$[W]^{t+1} = [W]^t + [\Delta W]^{t+1} \qquad (6.19)$$

- Find error rate as

$$err = \frac{\sum E_k}{nset} \qquad (6.20)$$

## 6.3 Face detection networks

The neural network-based face detection system operates in two stages. In the first stage it applies a set of neural network-based detectors to an image, and then during the second stage, it uses an arbitrator to combine the outputs. The individual detectors examine each location in the image at several scales, looking for locations that might contain a face. The arbitrator then merges detections from individual networks and eliminates overlapping detections.

The first component of this system is a neural network that receives as input a 20 × 20 pixel region of the image, and generates an output ranging from 1 to -1, signifying the presence or absence of a face, respectively. To detect faces anywhere in the input, the network is applied at every location in the image. To detect faces larger than the window size, the input image is repeatedly reduced in size (by subsampling), and the detector is applied at each size.

This network must have some invariance to position and scale. The amount of invariance determines the number of scales and positions at which it must be applied. In [110] the network at every pixel position in the image is applied, and the image is scaled down by a factor of 1.2 for each step in the pyramid.

After a window of certain size is extracted from a particular location, it is preprocessed using the affine lighting correction and histogram equalization steps. The preprocessed window is then passed to a neural network. Shapes of the subregions are chosen to allow the hidden units to detect local features that might be important for face detection. In particular, the horizontal stripes allow the hidden units to detect such features as mouths or pairs of eyes, while the hidden units with square receptive fields might detect features such as individual eyes, the nose or corners of the mouth. Other experiments have shown that the exact shapes of these regions do not matter; however it is important that the input is broken into smaller pieces instead of using complete connections to the entire input.

## 6.4 Training images

### 6.4.1 Data preparation

The first step in reducing the amount of variation between images of faces is to align the faces with one another. This alignment is necessary to reduce the variation in the two-dimensional position, orientation and scale of the faces. Ideally, the alignment would be computed directly from the images, using image registration techniques. Training of a detector is an important part, which can be done by creating new example images from real images. In [110], this has taken the form of randomly rotating, translating and scaling example images by small amounts. Once the faces are aligned to have a known size, position and orientation, the amount of variation in the training data can be controlled.

After aligning the faces and replacing the background pixels with more realistic values, variations due to lighting and camera characteristics are taken care of by a preprocessing technique. To equalize the intensity values across the window, a function is fitted which varies linearly across the window to the intensity values in an oval region inside the window (shown in Figure 6.3). Pixels outside the oval may represent the background, so those intensity values are ignored in computing the lighting variation across the face. If the intensity of a pixel $x, y$ is $I(x, y)$, then we want to fit this linear model parameterized by $a, b, c$ to the image:

$$
\begin{pmatrix} x & y & 1 \end{pmatrix} \cdot \begin{pmatrix} a \\ b \\ c \end{pmatrix} = I(x, y) \tag{6.21}
$$

By fitting a single linear plane to the image, unidirectional lighting effects can be corrected. This plane is computed efficiently through simple linear projection by solving the equation $[XY1] * C = I$ (where $X, Y$ and $I$ are the vectors corresponding to their respective coordinate values, 1 is a vector of 1's to compute the constant offset and $C$ is a vector of three numbers defining the linear slopes in the $X$ and $Y$ directions and the constant offset). To compute $C$, one simply needs to compute

$$
C = [(X \quad Y \quad O)^T (X \quad Y \quad O)]^{-1} (X \quad Y \quad O)^T I \tag{6.22}
$$

where $O$ contains the vector of 1's.

These plane coefficients in $C$ approximate the average gray level across the image under a linear constraint and thus can be used to construct a shading plane that can be subtracted out of the original image. Once the lighting direction is corrected for, the gray-scale histogram can then be rescaled to span the min and maximum gray-scale levels allowed by the representation. Next, histogram equalization is performed, which non-linearly maps the intensity values to expand the range of intensities in the window. The histogram is computed for pixels inside an oval region in the window. This compensates for differences in camera input gains, as well as improves contrast in some cases. Figure 6.3(b) shows the heavy directional lighting effects from left corresponding to the original image Figure 6.3(a).

Given the images in Figure 6.3(a) and Figure 6.3(b), we can subtract Figure 6.3(b) from Figure 6.3(a) and rescale the gray levels to the minimum and maximum range for our representation. We can then apply a mask to this image to remove background interference. This result is shown Figure 6.3(c).

In Figure 6.3, the unidirectional lighting effects have now been removed and the image in Figure 6.3(c) has approximately the same gray level distribution. This normalization is extremely important to proper functioning of the neural network.

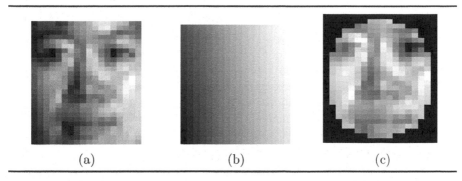

(a)  (b)  (c)

**FIGURE 6.3**: (a) Original image, (b) shading approximations, (c) normalized and masked image

### 6.4.2  Face training

In order to use a neural network to classify windows as faces or nonfaces, some training examples for each set are needed. The first step in reducing the amount of variation between images of faces is to align the faces with one another. This alignment should reduce the variation in the two-dimensional position, orientation and scale of the faces. Ideally, the alignment would be computed directly from the images, using image registration techniques. This would give the most compact space of images of faces. However, the image intensities of faces can differ quite dramatically, which would make some faces hard to align with each other, but every face should be aligned with every other face and this problem is solved in [110] by the manually labelling method. The algorithm for aligning manually labelled face images is as follows:

1. Initialize $F$ a vector which will be the average position of each labelled feature over all the faces, with some initial feature locations. In the case of aligning frontal faces, these features might be the desired positions of the two eyes in the input window. For faces of another pose, these positions might be derived from a 3D model of an average head.

2. For each face $i$, use the alignment procedure [110] to compute the best rotation, translation and scaling to align the face features $F_i$ with the average feature locations $F$. Call the aligned feature locations a $F_i'$.

3. Update $F$ by averaging the aligned feature locations $F_i'$ for each face $i$.

4. The feature coordinates in $F$ are rotated, translated and scaled to best

match some standardized coordinates [1]. These standard coordinates are the ones used as initial values for $F$.

5. Go to step 2.

After alignment, the faces are scaled to a uniform size, position and orientation within a proper window. The images are scaled by a random factor and also translated by a random amount up to half of a pixel. This allows the detector to be applied at each pixel location and at each scale in the image pyramid, and still detect faces at intermediate locations or scales. In addition, to give the detector some added invariance to variations in the faces, they are rotated by some specific amount.

A large number of nonface images are required to train the face detector, because the variety of nonface images is much greater than the variety of face images. Practically any image can serve as a nonface image and the space of nonface images is much larger than the space of face images. However, one should train the neural networks on precisely the same distribution of images. Training on a database of large size is very difficult. The next section describes approaches to training with a large amount of data.

### 6.4.2.1   Active learning

Active learning of the network is used, because of the difficulty of training with every possible negative example [110] and utilizes an algorithm described in [111]. Instead of collecting the images before training is started, the images are collected during training, in the following manner:

- Create an initial set of nonface images by generating 1000 random images. Apply the preprocessing steps to each of these images.

- Train a neural network to produce an output of 1 for the face examples, and -1 for the nonface examples. On the first iteration of this loop, the network's weights are initialized randomly. After the first iteration, the weights are computed by training in the previous iteration as the starting point.

- Run the system on an image of scenery which contains no faces. Collect subimages in which the network incorrectly identifies a face (an output activation $> 0$).

- Select some of these subimages at random, apply the preprocessing steps and add them into the training set as negative examples. Go to step 2.

The training algorithm used in step 2 is the standard error backpropagation algorithm with a momentum term. The neurons use the *tanh* activation function, which gives an output ranging from -1 to 1, hence

---

[1]Detailed description of alignment process is given in [110].

the threshold of 0 for the detection of a face. Since all the negative examples are not trained, the probabilistic arguments of the previous section do not apply for setting the detection threshold. Since the number of negative examples is much larger than the number of positive examples, uniformly sampled batches of training examples would often contain only negative examples, which would be inappropriate for neural network training. Instead, each batch of 100 positive and negative examples is drawn randomly from the entire training sets and passed to the backpropagation algorithm as a batch. The training batches have been chosen in a way such that they have 50% positive examples and 50% negative examples.

### 6.4.3   Exhaustive training

Neural network training usually requires training the network many times on its training images. This not only requires a huge amount of storage, but also takes many long hours. Additionally, a network usually trains on images in batches and consequently it may have forgotten the characteristics of the first image. To insure that the neural network learns about both faces and nonfaces, an exhaustive training of the batches of negative examples is made of approximately equal numbers of positive examples. However, this change may not support a real life situation of distribution. It is possible to compensate for this using the Bayes theorem by denoting $P(face|w)$ as the probability that a given window is a face, and $P(face)$ and $P(nonface)$ as the prior probability of faces and nonfaces in the training sets (both 0.5). Then probability of a face image training is given by

$$P(face|w) = \frac{P(w|face)P(face)}{P(w|face)P(face) + P(w|nonface)P(nonface)} \quad (6.23)$$

when $P(face|w)$ can be considered as the neural network output given by $NNout = P(face|w)$.

Neural networks will learn to estimate the left-hand side of this equation as $P(face), P(nonface)$. Since $P(w|nonface) = 1 - P(w|face)$, this equation is simplified as

$$P(w|face) = NNout \quad (6.24)$$

Let us denote the probability of the presence of a face (true probability) as $P(face)$, and that nonface is $P(nonface)$. Then using Bayes' theorem the true probability for the presence of a face in the window is given by

$$P(face|w) = \frac{NNOutput * P(face)}{NNoutP(face) + (1NNout)P(nonface)} \quad (6.25)$$

The window is classified as face if $P(face|w) > 0.5$, which is equivalent to setting a threshold as $NNout > 1P(face)$. Since we are using neural networks with *tanh* activation functions, the output range is $\pm 1$.

The outputs from the face detection networks are not binary. The neural networks produce real values between 1 and -1, indicating whether or not the input contains a face. A threshold value of zero is used during training to select the negative examples. To examine the effect of this threshold value during testing, the false positive rate is measured, as the threshold is varied from 1 to -1. At a threshold of 1, the false detection rate is zero, but no faces are detected. As the threshold is decreased, the number of correct detections increases, but so do the number of false detections.

---

## 6.5  Evaluation of face detection for upright faces

### 6.5.1  Algorithm

To achieve face detection using a neural network, the following general algorithm can be used.

1. Normalize training data for each face and non-face image.

   (a) Subtract out an approximation of the shading plane to correct for single light source effects.

   (b) Rescale histograms so that every image has the same gray level range.

   (c) Aggregate data into labeled data sets.

2. Train the neural net

   (a) Until the neural net reaches convergence or a decrease in performance on the validation set.

   (b) Perform gradient descent error backpropagation on the neural net for the batch of all training data.

3. Apply the face detector to image.

   (a) Build a resolution pyramid of the image by successively decreasing the image resolution at each level of the pyramid, stopping at some default minimum resolution.

   (b) For each level of the pyramid

      i. Scan over the image, applying the trained neural net face detector to each rectangle within the image.

      ii. If a positive face classification is found for a rectangle, scale this rectangle to the size appropriate for the original image and add it to the face bounding-box set.

4. Return the rectangles in the face bounding-box set.

## 6.5.2 Image scanning and face detection

For a constant size input and constant size mask, some method is needed for scaling the image so that it can detect faces of multiple sizes. Consequently, an image pyramid is built for an image to be scanned by placing the original image at the bottom and successively scaling down the resolution between pyramid levels until some preset low resolution level has been reached. An example of an image pyramid having six levels and a scale factor of 1.2 is shown Figure 6.4.

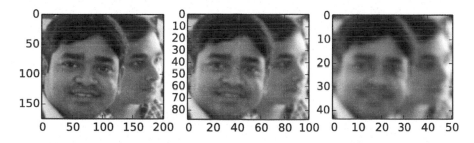

**FIGURE 6.4**: Images at different scales with scale factor 1.2 are shown

Once the image pyramid is obtained, face detection is a straightforward process. For each level of the image pyramid, scanning is performed over all possible rectangles. Then each rectangle is extracted from the image and normalized. Then the neural network is used for classification. The neural net returns a value which can be thresholded to determine whether that image is a face or not. It is fairly straightforward to compute the bounding box of the face for the original scaling from the image level and location of the rectangle. Consequently, all face bounding-boxes are stored and passed back to the calling procedure. This set of bounding boxes outlines the predicted face locations and scales in the image and can be overlaid on the image.

Figure 6.5 shows that some face detection result. Moreover the test images taken here are mostly frontal. Rotation invariant face detection, however, needs more refinement of the network and very exhaustive training of the network.

## MATLAB code for shading correction

```
%% Shading correction

clear all
close all
clc

im = imread("....jpg");
im = double(im);
IN{1}  = imresize(im,[27,18]);
imshow(IN{1},[]);
MASK = buildmask;

% Retrieve the indices for the given mask
IND = find(MASK);
figure
% Set up matrices for planar projection calculation
% i.e. Ax = B  so  x = (transpose(A)*A)^-1 * transpose(A)*B
x = 1:1:size(IN{1},2);
y = 1:1:size(IN{1},1);
[mx,my] = meshgrid(x,y);
mxc = mx(IND);
myc = my(IND);
mcc = ones(size(myc));
A = [mxc, myc, mcc];

% Cycle through each image removing shading plane
% and adjusting histogram
for i=1:1

    % Calculate plane: z = ax + by + c
    B = IN{i}(IND);
    x = inv(transpose(A)*A)*transpose(A)*B;
    a = x(1); b = x(2); c = x(3);

    %This is the color plane itself
    SHADING{i} = mx.*a + my.*b + c;
    imshow(SHADING{1},[])
    %This is the image minus the color plane
    %(the constant will be normalized out in histogram recentering)
    OUT{i} = IN{i} - (mx.*a + my.*b + c);

    % Now, recenter the histogram
    maximum = max(max(OUT{i}.*MASK));
    minimum = min(min(OUT{i}.*MASK));    %minimum = min(min(OUT{i}))
    diff = maximum - minimum;
    OUT{i} = ((OUT{i}-minimum)./diff).*MASK;
```

```
end
figure
imshow(OUT{1},[]);
Histout = histeq(OUT{1});
figure
imshow(Histout,[])
```

**FIGURE 6.5**: Multiple face detection results using the neural network approach.

# Chapter 7

## Real-time face detection

## 7.1  Introduction

A real-time face detection framework works in situations where the processing of face images is carried out extremely rapidly with high true detection rates. The most prominent system which achieves such an objective is known as the Viola-Jones face detector [112],[113] and the material of this chapter is mainly based on his works. In the majority of cases the technique is implemented in wide ranges of small,low, power devices, including handheld and embedded processors. This extremely fast face detector has broad practical applications where rapid frame-rates are not necessary. These include user interfaces, image databases and teleconferencing. Though the training is slow in this system, the detection rate is very fast. This face detector has three main ideas: (1) new image representation called integral image allows for very fast feature evaluation, (2) a simple and efficient classifier that is built by selecting a small number of important features from a huge library of potential features using an algorithm called AdaBoost and (3) a method for combining successively more complex classifiers in a cascade structure which dramatically increases the speed of the detector by focusing attention on promising regions of the image.

## 7.2    Features

The Viola-Jones face detection procedure classifies images based on the value of simple features. There are many motivations for using features rather than the pixels directly. The most common reason is that features can act to encode adhoc domain knowledge that is difficult to learn using a finite quantity of training data. The second critical motivation for accessing features is related to the notion that the feature-based system operates much faster than a pixel-based system.

The Viola-Jones algorithm uses Haar-like features, that is, a scalar product between the image and some Haar-like templates. More specifically, it uses three kinds of features :1) two-rectangle feature, 2) three-rectangle feature and 3) four-rectangle feature. The value of a two-rectangle feature is the difference between the sum of the pixels within two rectangular regions. A three-rectangle feature computes the sum within two outside rectangles subtracted from the sum in a center rectangle. Finally a four-rectangle feature computes the difference between diagonal pairs of rectangles. The regions have the same size and shape and are horizontally or vertically adjacent. Five Haar-like patterns are shown in Figure 7.1.

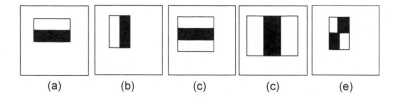

    (a)         (b)         (c)         (c)         (e)

**FIGURE 7.1**: Example rectangle features shown relative to the enclosing detection window. The sum of the pixels which lie within the white rectangles is subtracted from the sum of pixels in the grey rectangles. Two-rectangle features are shown in (a) and (b), three-rectangle features are shown in (c) and (d) and the four-rectangle feature is shown in (e)

The size and position of a pattern's support can vary provided its black and white rectangles have the same dimension, border each other and keep their relative positions. Let $I$ and $P$ denote an image and a pattern, both of the same size $N \times N$ (as Figure 7.2). The feature associated with pattern $P$ of image $X$ is defined by

$$\sum_{i=1}^{N}\sum_{j=1}^{N} X(i,j)1_{P(i,j)white} - \sum_{i=1}^{N}\sum_{j=1}^{N} X(i,j)1_{P(i,j)black} \qquad (7.1)$$

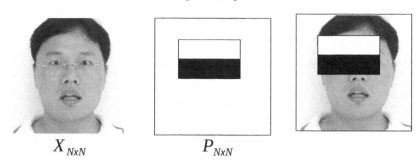

**FIGURE 7.2**: Haar-like features. Only those pixels marked in black or white are used when the corresponding feature is calculated.

---

## 7.3 Integral Image

Rectangular features can be computed very rapidly using an intermediate representation for the image which is called the integral image. The integral image can be computed in one pass over the original image as shown in Figure 7.3. The integral image at location $(i, j)$ contains the sum of the pixels above and to the left of $(i, j)$, and is given by

$$II(i, j) = \sum_{i \leq i', j \leq j'} I(i, j) \qquad (7.2)$$

where $II(i, j)$ is the integral image and $I(i, j)$ is the original image.

Using the following pair of recurrences:

$$
\begin{aligned}
S(i, j) &= S(i, j - 1) + I(i, j) \\
II(i, j) &= II(i - 1, j) + S(i, j)
\end{aligned}
\qquad (7.3)
$$

where $S(i, j)$ is the cumulative row sum and $II(-1, j) = 0$ and $S(i, -1) = 0$.

Using the integral image any rectangular sum can be computed in four array references as shown in Figure 7.4, which gives the idea of how to evaluate the sum of pixels of the original image inside the rectangle D, which can be computed with four array references. The value of the integral image at location 1 is the sum of the pixels in rectangle A. The value at location 2 is A + B, at location 3 is A + C and at location 4 is A + B + C + D. The sum within D can be computed as 4 + 1 (2 + 3). More detailed explanation is given in the next section and in Figure 7.5.

### 7.3.1 Rectangular feature calculation from integral image

Two-rectangular feature calculation is shown in Figure 7.5, where a two-rectangle feature of size $2 \times 2$ is considered and overlaid on the integral array,

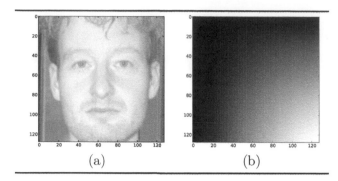

(a)                          (b)

**FIGURE 7.3**: (a) Original image, (b) its integral image.

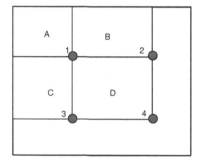

**FIGURE 7.4**: The sum of pixels of the original image inside the rectangle D can be computed with four array references

according to the integral image of the white portion $(a - c - d1 - b1)$ of the rectangle feature, point $a = 5, b1 = 7, c = 8$ and $d1 = 17$. For the rectangle $(a-c-d1-b1)$, the sum is calculated as $d1+a-b1-c1 = 17+5-7-8 = 7$ which is equal to $5 + 2 = 7$ in the original array. Inside the black area the rectangle sum is also calculated. The black rectangle portion $(b1 - d1 - d2 - b2)$ of the two-rectangle feature, $d2 = 25, b2 = 10, b1 = 7, d1 = 17$, and the rectangle sum of this black area are calculated as $d2+b1-b2-d1 = 25+7-10-17 = 5$, which is equal to $4 + 1 = 5$ in the original image. Hence, for the two-rectangle feature, only six array references (as $5, 7, 10, 8, 17, 25$) are needed. Eight array references are obtained in the case of the three-rectangle features and nine for four-rectangle features. These meaningful sets of rectangle features have the property that a single feature can be evaluated at any scale and location in a few operations.

Effective face detectors can be constructed with as few as two rectangle features. Given the computational efficiency of these features, the face detection process can be completed for an entire image at high speed. Even though each feature can be computed very efficiently, computing the complete set is prohibitively expensive. This problem is handled by forming an effective

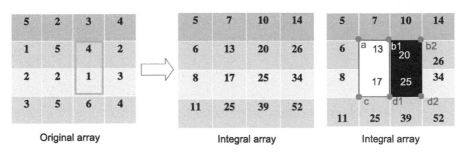

**FIGURE 7.5**: Two-rectangle feature calculation from an integral array

classifier by combining these features. The main challenge is to find these features. In the Viola-Jones face detection method, a variant of AdaBoost is used both for selecting the features and to train the classifier.

The AdaBoost is a learning algorithm and is used to boost the classification performance of a simple learning algorithm, by combining a collection of weak classification functions to form a stronger classifier. In the language of boosting, the simple learning algorithm is called a weak learner as we do not expect even the best classification function to classify the training data well. In order to boost the weak learner, it is called upon to solve a sequence of learning problems. After the first round of learning, the examples are reweighted in order to emphasize those which were incorrectly classified by the previous weak classifier. The final strong classifier takes the form of a weighted combination of weak classifiers followed by a threshold.

The conventional AdaBoost procedure can be easily interpreted as a greedy feature selection process. In the case of a general problem of boosting a large set of classification functions are combined using a weighted majority vote. The challenge is to associate a large weight with each good classification function and a smaller weight with poor functions. AdaBoost is also an aggressive mechanism for selecting a small set of good classification functions which nevertheless have significant variety. Drawing an analogy between weak classifiers and features, AdaBoost is an effective procedure for searching out a small number of good features which have significant variety.

---

## 7.4 AdaBoost

As has been said, AdaBoost is an algorithm for constructing a strong classifier as a linear combination of simple weak classifiers. Final classification is based on the weighted vote of weak classifiers. Pseudocode for AdaBoost is given in the following.

Given $(x_1, y_1), ..., (x_m, y_m)$ where $x \in X$ and $y \in Y = \{+1, -1\}$,
Initialize $D_1(i) = 1/m$
For $t = 1, ..., T$ :

- Train weak learner using distribution $D_t$

- Get weak hypothesis : $h_t : X \to \{-1, +1\}$ with error

$$\epsilon_t = Pr_i{}_{D_t}[h_t(x_i) \neq y_i]$$

- Choose

$$\alpha_t = 1/2ln\left(\frac{1 - \epsilon_t}{\epsilon_t}\right)$$

- Update

$$
\begin{aligned}
D_{t+1}(i) &= \frac{D_t(i)}{Z_t} \times e^{-\alpha_t} \quad if \quad h_t(x_i) = y_i \\
&= \frac{D_t(i)}{Z_t} \times e^{\alpha_t} \quad if \quad h_t(x_i) \neq y_i \\
&= \frac{D_t(i)exp(-\alpha_t y_i h_t(x_i))}{Z_t}
\end{aligned}
$$

(7.4)

where $Z_t$ is a normalization factor (chosen so that $D_{t+1}$ is a distribution).

Output of the final hypothesis is given by

$$H(x)sign\left(\sum_{t=1}^{T} \alpha_t h_t(x)\right) \tag{7.5}$$

The algorithm takes as input a training set $(x_1; y_1); ...; (x_m; y_m)$ where each $x_i$ belongs to some domain or instance space $X$ and each label $y_i$ is in some label set $Y$. AdaBoost calls a given weak or base learning algorithm repeatedly in a series of rounds $t = 1, ..., T$. One of the main ideas of the algorithm is to maintain a distribution or set of weights over the training set. The weight of this distribution on training example $i$ on round $t$ is denoted $D_t(i)$. Initially, all weights are set equally, but on each round, the weights of incorrectly classified examples are increased so that the weak learner is forced to focus on the hard examples in the training set. The weak learner's job is to find a weak hypothesis $h_t : X \to \{-1, +1\}$ appropriate for the distribution $D_t$.

## 7.4.1 Modified AdaBoost algorithm

Among all the features to be accessed, some are expected to give almost consistently high values particularly when on top of a face. In order to find these features Viola-Jones uses a modified version of the AdaBoost algorithm as given below.

Given the numbers of example images $(x_1, y_1), ..., (x_n, y_n)$ where $y_1 = 0, 1$ for negative and positive examples,
Initialize weights $w_{1,i} = \frac{1}{2m}, \frac{1}{2l}$ for $y_1 = 0, 1$, where $m$ and $l$ are positive and negative examples.
For $t = 1, \cdots, T$ :

- Normalize the weights:

$$w_{t,i} \leftarrow \frac{w_{t,i}}{\sum\limits_{j=1}^{n} w_{t,j}}$$

- Select the best weak classifier with respect to the weighted error:

$$\epsilon_t = \min_{f,p,\theta} \sum_i w_i |h(x_i, f, p, \theta) - y_i|$$

- Define $h_t(x) = h(x, f_t, p_t, \theta_t)$ where $f_t, p_t$ and $\theta_t$ are the minimizers of $\epsilon_t$.

- Update the weights:

$$w_{t+1,i} = w_{t,i} \beta^{1-e_i}$$

where $e_i = 0$ if example $x_i$ is classified correctly and $e_i = 1$ otherwise, and $\beta_t = \frac{\epsilon_t}{1-\epsilon_t}$.

- The final strong classifier is:

$$C(x) = \begin{cases} 1 & \text{if } \sum\limits_{t=1}^{T} \alpha_t h_t(x) \geq \frac{1}{2} \sum\limits_{t=1}^{T} \alpha_t; \\ 0 & otherwise \end{cases}$$

where $\alpha_t = log \frac{1}{\beta_t}$.

An important part of the modified AdaBoost algorithm is concerned with the determination of the best feature, polarity and threshold. To achieve a smart solution to this problem Viola-Jones suggested a simple brute force method. This means that the determination of each new weak classifier involves the evaluation of each feature on all the training examples. The best performing feature is chosen based on the weighted error it produces. In the modified AdaBoost algorithm, the weight of a correctly classified example is decreased while the weight of a misclassified example is kept constant. As a result it is more expensive for the second feature (in the final classifier) to misclassify an example. Thus the second feature is forced to focus harder on the examples misclassified by the first. The point is that the weights are a vital part of the mechanics of the AdaBoost algorithm. Furthermore, the final classifier is a weighted sum of the weak classifiers. It is called weak because it alone can not classify the image, but together with others forms a strong classifier.

---

PYTHON Code for AdaBoost classification

---

```python
# AdaBoost classification example

from numpy import *

x= array([[0,1],[1,1],[2,1],[3,-1],[4,-1],\
[5,-1],[6,1],[7,1],[8,1],[9,-1]])
p = array([[0,0.1],[1,0.1],[2,0.1],[3,0.1],\
[4,0.1],[5,0.1],[6,0.1],[7,0.1],[8,0.1],[9,0.1]])
h_final =0
Thres = zeros((4,1))
alpha=zeros((4,1))

for t in range(0,3):
    err= zeros((9,1))
    thr = array([0.5,1.5,2.5,3.5,4.5,5.5,6.5,7.5,8.5])

    for k in range(0,9):
            if t ==2:
                h = sign(x[:,0]-thr[k])
            else:
                h = sign(thr[k]-x[:,0])

            for j in range(0,10):
                if h[j] != x[j,1]:
                    err[k] = err[k] + p[j,1]

    for l in range(0,9):
        if err[l] == err.min():
            indx = l
            break
    Thres[t]= thr[l]

    if t==2:
        h = sign(x[:,0]-thr[l])
    else:
        h = sign(thr[l]-x[:,0])

    alpha[t] = 0.5 * log((1-err.min())/err.min())
```

```
q1 = exp(-alpha[t])
q2 = exp(alpha[t])
Zt = 2*sqrt(err.min()*(1-err.min()))


for j in range(0,10):
    if h[j] == x[j,1]:
        p[j,1] = (q1*p[j,1])/Zt
    else:
        p[j,1] = (q2*p[j,1])/Zt


f = alpha[t]*(h)
h_final = h_final+f

decision = sign(h_final)
print decision
```

---

## 7.4.2   Cascade classifier

This part describes an algorithm for constructing a cascade of classifiers which achieves increased detection performance while radically reducing computation time. Smaller, and therefore more efficient, boosted classifiers can be constructed which reject many of the negative sub-windows while detecting almost all positive instances. Simpler classifiers are used to reject the majority of sub-windows before more complex classifiers are called upon to achieve low false positive rates.

The basic principle of the Viola-Jones face detection algorithm is to scan the detector many times through the same image, each time with a new size. Even if an image should contain one or more faces it is obvious that an excessive large amount of the evaluated sub-windows would still be negatives (non-faces). This realization leads to a different formulation of the problem. Instead of finding faces, the algorithm should discard non-faces. It is faster to discard a non-face than to find a face. Therefore, a detector consisting of only one (strong) classifier suddenly seems to be inefficient since the evaluation time is constant no matter how the input is. Hence the need for a cascaded classifier arises.

The cascaded classifier is composed of stages, each containing a strong classifier. The job of each stage is to determine whether a given sub-window is definitely not a face or maybe a face. When a sub-window is classified to be a non-face by a given stage it is immediately discarded. Conversely a sub-window classified as a maybe-face is passed on to the next stage in the cascade. It follows that the more stages a given sub-window passes, the higher is the chance of a sub-window in containing a face. The concept is illustrated with two stages in Figure 7.6. In a single stage classifier, one

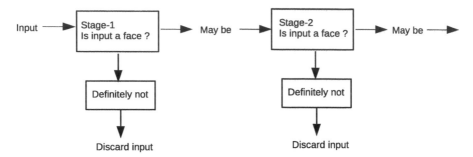

**FIGURE 7.6**: Cascade classifier structure

would normally accept false negatives in order to reduce the false positive rate. However, for the first stage in the staged classifier, false positives are not considered since the succeeding stages are expected to sort them out. Therefore, many false positives in the initial stages are accepted. Consequently the amount of false negatives in the final stage of the classifier is expected to be very small. Therefore, more attention (computing power) is directed towards the regions of the image suspected to contain faces. The training algorithm for building a cascade detector is given as follows:

- User selects values for $f$, the maximum acceptable false positive rate per layer, and $d$, the minimum acceptable detection rate per layer

- User selects target overall false positive rate, $F_{target}$

- P = set of positive examples

- N = set of negative examples

- $F_0 = 1.0; D_0 = 1.0$

- $i = 0$

- while $F_i > F_{target}$ $i \leftarrow i + 1$
  $n_i = 0; F_i = F_{i-1}$
  while $F_i > f \times F_{i-1}$


- $n_1 \leftarrow n_i + 1$

- Use P and N to train a classifier with $n_i$ features using AdaBoost

- Evaluate current cascaded classifier on validation set to determine $F_i$ and $D_i$

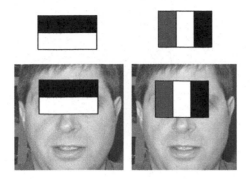

**FIGURE 7.7**: Two features obtained as the best features from AdaBoost

- Decrease threshold for the $i$th classifier until the current cascaded classifier has a detection rate of at least $d \times D_{i-1}$ (this also affects $F_i$)

- $N \leftarrow \phi$

- If $F_i > F_{target}$ then evaluate the current cascaded detector on the set of non-face images and put any false detections into the set $N$

Introducing the cascade classifier instead of applying all the features on a window, it may be necessary to group the features into different stages of classifiers and apply one-by-one. Normally the first few stages contain a much less number of features. If a window fails during the first stage of operation, it is discarded. If it passes, the second stage is continued. The window which passes all stages indicates the presence of a face region. Two features shown in Figure 7.7 are actually obtained as the best two features from AdaBoost.

## 7.5 Face detection using OpenCV

OpenCV comes with a trainer as well as detector. These programs are freely available in www.opencv.com. OpenCV contains many pre-trained classifiers for faces. The XML file for Haar-cascade face training is stored in the opencv/data/haarcascades/ folder. To create a face detector with OpenCV, we need to load the required XML classifiers and then load any input image (or video) in gray-scale mode. Then the program attempts to find the faces in the image. If faces are found, it returns the positions of detected faces. Once these locations are obtained, a region of interest (ROI) for the face is established.

Figure 7.8 shows some examples of face detection using OpenCV Haar-cascade detection. Multiple face detection is shown in Figure 7.8(a) and (b).

(a)                    (b)

**FIGURE 7.8**: Multiple face detections are shown in (a) and (b) by varying the threshold values.

The faces that are not detected in (a) are recognized as faces in (b) by simply changing the threshold.

PYTHON Code for OpenCV based face detection[1]

```
# OpenCV based face detection
import numpy as np
import cv2
face_cascade = cv2.CascadeClassifier \
("haarcascade_frontalface_alt2.xml")
img = cv2.imread("/home/pradipta/PRADIPTA/" \
"Database/CalTechfaces/16.jpg",0)
faces = face_cascade.detectMultiScale(img, 1.3,5)
for (x,y,w,h) in faces:
    cv2.rectangle(img,(x,y),(x+w,y+h),(255,0,0),2)
    roi_gray = img[y:y+h, x:x+w]
cv2.imshow("Faces",img)
```

[1] http://docs.opencv.org/trunk/doc/py-tutorials/py-objdetect/py-face-detection/py-face-detection.html

# Chapter 8

# Face space boundary selection for face detection and recognition

## 8.1 Introduction

A face recognition task can be viewed in general, as a combination of two stages, i.e., the face detection or verification stage and the face identification stage. The face detection algorithms usually include a verification phase at the decision level to discriminate the face and non-face portions of an image. These types of face detection problems can be tackled as a classification problem, where the given query image is classified into a face image (if a face is present in it) or a non-face image (if no face is present in it). For the purpose of classification, generally one needs to have samples from both the classes, and hence, as a training set, collections of face images and of non face images are required. The basic problem in this setup is to have information about all face and non face images. Division of the face space into subclasses can be done using a framework of higher order statistics to model the face and non-face clusters. Methods for face detection were also proposed [114] which seek to represent the wider variety of human faces as a set of subclasses.

The classical approach of identification is termed as closed test identification, where the test face image always exists in the client database. However, in a real life scenario the identification system may be put in a situation, where no face image corresponding to the person in the query face image is present in the database. This case is often referred to as the open test identification, where the system is required to identify the test face image as an imposter to the system. A way of achieving the open test identification task is to put a threshold on the dissimilarity value at the identification stage. Thus in conventional face recognition on the basis of the decision threshold, the recognition system should be in a position to detect a face and then accept the query image as client or reject the face image as impostor.

For finding the face space boundary, or in other words, for finding a decision threshold for face space, the standard biometric technique based on receiver operating characteristics (ROC) is utilized. ROC employs the false acceptance rate (FAR) curve and false rejection rate (FRR) curve to obtain the threshold, which is generally selected corresponding to the equal error rate (EER). However, it is extremely difficult to obtain a good estimate of FAR, if the given non face images are not the representatives of all non face classes. In reality, a system can have extremely few examples of genuine access and relatively few impostor accesses. As a result, user-specific threshold selection involving FAR and FRR may not always be reliable since reliable estimates of FAR may not be available. The common practice is to obtain a global threshold (for deciding a query to be a client or an imposter) for a system rather than calculating thresholds for each user using user specific ROCs. Selection of global, local or user-specific thresholds can give rise to different types of decisions. Global threshold provides a decision whether the query image is a face or non-face i.e. determines the detection of faces. The local threshold identifies the user as a member of one of the training classes in the database. These two thresholds will satisfy a biometric system in both for detection and identification cases.

## 8.2 Face points, face classes and face space boundaries

The face detection method considered here is an algorithm, wherein one class is a *face class* and the other is a *non-face class*. However, the training sample points are from the face class alone. This face class represents the space of *all* face images in the image space. Usually, the given images of faces are faces of several persons with variations in rotation, scaling, occlusions, expressions and viewing angles. Face images with artifacts and with light variations may also be present. The term *face class of a person* is used to indicate the set of all face images of all persons. Main characteristics of a face, like eyes, ears, mouth, etc., are reflected in face space. However, there are no such identifiable characteristics for the non-face class.

For good representation of the non-face class, one needs to use very large number of images reflecting different facets of the non-face class. This is an extremely difficult situation and any method employing ROC will encounter the problem of non reliability related to the false acceptance rate (FAR). Moreover, when subspace techniques are applied to obtain face space for dimensionality reduction, the utility of finding distance from the face space (DFFS) and distance in face space (DIFS) may be in question. Thus, instead of trying to find representative points for non-face class, one may try to obtain the boundary for face space and declare any point lying outside the boundary as non-face. In this procedure, training sample points from non-face class may not be required. Guided by this intuition, the set estimation technique for finding the global boundary or global threshold of face space is utilized.

The set estimation method can also be effective for face classification or the recognition problem. The numbers of classes are the same as the number of persons. $M$ images of human beings are considered and for each human being $N$ face images of same size and same background are also considered available. If an image with an expression of a person is represented by a vector $x_0$, then the set corresponding to the small variations in the same expression may be assumed to be a disc of radius $\epsilon > 0$ around $x_0$. The set corresponding to an expression of the same person may be taken as $\bigcup_{i=1}^{n}\{x \in \Re^m : d(x_i, x) \le \epsilon\}$, where $x_1, x_2, \cdots, x_n$ are the given $n$ vectors corresponding to the given $n$ images of the same expression for the same person. The dimension of the vectors is assumed to be $m$. The set corresponding to the union of all possible expressions of a person may also be taken as a connected set. The face class of a person is then nothing but the set of all possible expressions of that person. The radius value is taken to be independent of the center of the disc. Moreover, as the number of face images of the same person increases, more information regarding the face class is available and hence the radius value needs to be decreased. Thus, the radius value is a function of the number of images and each face class has an intraclass local threshold which can determine the class boundaries.

## 8.3    Mathematical preliminaries for set estimation method

Set estimation basically refers to the reconstruction of an unknown set from a random sample of points whose distribution is related to it. The problem has been addressed in different fields of research. The method of set estimation is mainly used to find the pattern class and its multivalued shape/boundary from its sample points in the two dimensional feature space $\Re^2$ [115]. It was found to be useful in developing a multivalued recognition system. The problem is related to estimation of an unknown set $\alpha$ from a random sample of points $S_n = \{X_1, X_2, ..., X_n\}$. Here $X_i$'s are independent and identically distributed random vectors, the support of each $X_i$ is $\alpha$, and the probability density function on $\alpha$ is assumed to be continuous. In computational geometry, for instance, the efficient construction of convex hulls for finite sets of points has important applications in pattern recognition, cluster analysis and image processing, among others.

Estimator for a support set was evolved in 1964 [116], [117] and was later modified. A case was studied where $\alpha$ is a convex support in the two-dimensional Euclidean space. This led to the development of a natural estimator, the convex hull of the sample $S_n$. However, if $\alpha$ is not convex, the convex hull of the sample is not an appropriate estimator. The estimate of a set is difficult if no assumptions are made on its shape. In this setting, it was proposed [118] to estimate the support of an unknown probability measure by means of a smoothed version of the sample $S_n$. Results on the performance of the estimator were later obtained and refined recently [119]. Investigations on estimation of $\alpha$-shapes for point sets in $\Re^3$ had been studied and was extended [120] to $\Re^m$. As one can get the shape or boundary of a given set, the procedure of set estimation also determine the class thresholds of the set. As a tool of set estimation, minimal spanning tree (MST) is used to calculate threshold value. The minimal spanning tree (MST) is the spanning tree of minimum length. Given a connected, undirected graph, a spanning tree of the graph is a subgraph which connects all the vertices together. More generally, any undirected graph (not necessarily connected) has a tree, which is an union of spanning trees for its connected components. Many algorithms are available for obtaining an MST.

In order to apply the set estimation procedure, one needs to know whether the assumed properties of sets in the estimation procedure hold for the face space. It may be noted that the face class $\alpha$ of any person is assumed to be path connected and compact, that is $Cl(int(\alpha)) = \alpha$ and $\lambda(\delta\alpha) = 0$, where $cl$ denotes closure, and $Int$ denotes the interior. The other two main assumptions are path connectivity and the boundedness of the set. Fortunately, the face space $\alpha$ is bounded, since gray values are bounded. One may also assume that, theoretically at least, the face space is path connected, as between any two faces one can construct a path consisting of faces in which there are little variations between two consecutive face images. As a consequence, the set estimation procedure can be applied for determining the boundaries for each face class and the whole face space. Consequently recognition and identification of non-faces and face class is possible. However, several issues may arise out of the formulation which is mainly related to the value of $\varepsilon$. These issues were addressed by different authors [115],[121],[120]. In general, one may want to estimate a set on the basis of the given points with useful properties.

Some definitions may be needed in order to formulate the set estimation problem as a problem of finding a consistent estimate of a set [122]. The definitions are stated as:

**Definition 1**: $\alpha \subseteq \Re^m$ is said to be path connected, if for any two points $x, y \in \alpha, x \neq y, \exists$ a function $f : [0,1] \to \alpha$ such that $f$ is continuous, $f(0) = x$ and $f(1) = y$.

**Definition 2**: Let $X_1, X_2 \ldots, X_n \ldots$ be a sequence of independent and identically distributed random vectors which follow some continuous distribution over the set $\alpha \subseteq \Re^m$, where $\alpha$ is an unknown quantity. Let $\alpha_n$ be an estimated set based upon the random vectors $X_1, X_2 \ldots, X_n \ldots$. Then $\alpha_n$ is said to be a consistent estimate of $\alpha$, if $E_\alpha[\mu(\alpha_n \Delta \alpha)] \to 0$ as $n \to \infty$, where $\Delta$ denotes symmetric difference, $\mu$ is the Lebesgue measure [123] and $E_\alpha$ denotes the expectation taken under $\alpha$.

Let $X_1, X_2 \ldots, X_n \ldots$ be independent and identically random vectors, which follow uniform distribution over $\alpha \subseteq \Re^2$, where $\alpha$ is unknown. Let $\alpha$ be such that $cl(Int(\alpha)) = \alpha$ and $\mu(\delta\alpha = 0)$, where $\delta\alpha$ denotes the boundary of $\alpha$, $cl$ denotes closure, and $Int$ denotes the interior. Let $\{\epsilon_n\}$ be a sequence of positive numbers such that $\varepsilon_n \to 0$ and $n\varepsilon_n^2 \to \infty$, as $n \to \infty$. Let

$$\alpha_n = \bigcup_{i=1}^{n} \{x \in \Re^2 : d(x, X_i) \leq \epsilon_n\} \tag{8.1}$$

where $d$ denotes the Euclidean distance. As a consequence, $\alpha_n$ is a consistent estimate of $\alpha$.

A way of finding $\{\epsilon_n\}$ for points in two dimensional spaces is proposed by Murthy [124] by generalizing the method applicable to any continuous density function on $\alpha$, where $\alpha$ is a path-connected compact set. The essence of his method is as follows:

*i*) Find the minimal spanning tree (MST) $G_n$ of $S_n$, where the edge weight is taken to be the Euclidean distance between two points. Note that the MST of $S_n$ would be an uncountable set of points.

*ii*) Let $l_n$ denote the sum of edge weights of MST, $\varepsilon_n = \sqrt{\frac{l_n}{n}}$ and

*iii*) $E_n = \bigcup_{i=1}^{n} \{x \in \Re^2 : d(x, X_i) \leq \epsilon_n\}$ and that $E_n$ is a consistent estimate of $\alpha$.

The method provides a definition for $\epsilon_n$, and can be easily generalized to $m$ dimensions for any continuous probability density function $f$ on $\alpha$ satisfying the conditions: (1)for any open set $v \subseteq \Re^m, v \cap \alpha \neq \phi \Rightarrow \int_v f dx > 0$, and the parameter $\varepsilon_n$ is given by $\varepsilon_n = (l_n/n)^{1/m}; m \geq 2$ and (2) as $n \rightarrow\rightarrow \infty, \varepsilon_n \rightarrow 0$ and $n\varepsilon_n^m \rightarrow \infty$. The value of $m$, however, may change from dataset to dataset.

---

## 8.4    Face space boundary selection using set estimation

The face detection problem is considered here as a bi-class classification problem. Several face images from different face datasets are considered as a training set. Consider images of $M$ human beings, and for each human $N$ face images of the same size and same background are available. If an image is represented by a vector $\underline{x}$, total number of such images to form the face class is therefore $MN$ and these images are used for projection to the face space. Dimensionality reduction can be carried out by conventional subspace method, and the reduced dimension value is denoted by $m$.

As there are face images of $M$ persons in the training set, the estimated set corresponds to the face space of those $M$ face images. If a point is declared to be outside this estimated face boundary, then the point corresponds to either a non-face or a face image outside this face space (impostor) to the system. For large $M$, the estimated boundary is expected to be close to the actual, and thus any point outside the face space is expected to be a non-face.

### 8.4.1    Algorithm for global threshold-based face detection

The threshold to be calculated is called the global threshold, since the threshold is designed for the whole system and not for face images of a single person. For determining the global threshold, the following algorithmic steps are followed:

**Step 1:** Find minimum spanning tree (MST) of $MN$ points and find half of its maximal edge weight. Let it be denoted as the global threshold $\xi$.

**Step 2:** This is related to the process of detection. Let $\rho_i$ denotes the minimum distance between the given point and the points in the $i$ th class and let

$$\rho = min_{i=1,2,...n}\rho_i \tag{8.2}$$

If $\rho < \xi$, then the image is a face image. If $\rho > \xi$, then the given vector $x$ does not belong to the given face space and the image is either a non face image or a face image not belonging to the given face space (i.e., an impostor to the system). The pictorial description of the proposed global threshold is presented in Fig. 8.1.

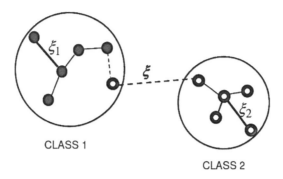

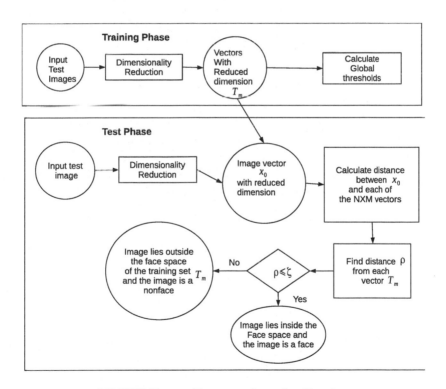

**FIGURE 8.1**: Global threshold selection in face space

**FIGURE 8.2**: Face non-face classification

The local thresholds $\xi_i$'s are calculated by considering that each $\xi_i$ represents the maximal edge weight of MST of the points in the training set for the $i$ th class. As the number of points in a class increases, $\xi_i$ decreases. If the number of persons is taken to be fixed and the number of images per person increases, then $\xi_i$ is very small. In such a situation, $\xi$ is greater than each $\xi_i$. $\xi$ is selected as the global threshold for the system. In many cases, however, $\xi$ is found to be greater than each $\xi_i$, though it does not necessarily happen always for every $M$ and $N$. On the other hand, if $M$ is increased keeping $N$ constant, the values of some of the $\xi_i$'s are expected to be greater than $\xi$. The block diagram of the suggested method of thresholding as applied to face and non-face detection is shown in Figure 8.2.

---

## 8.5     Experimental design and result analysis

For the detection of face / non face, the experiment can be carried out in many ways. One way is to have images which contain either no face or exactly a single face without any background. The other way is to have multiple faces with few skin regions (hand, neck, leg portions, etc.) and with backgrounds.

### 8.5.1     Face/non-face classification using global threshold during face detection

A way of conducting experiments for face / non-face classification is to take some face images, estimate the boundary of the face space, and check whether any non face image belongs to the estimated face space. For this purpose, the COIL-100 [125] dataset is used, where the color images are converted into gray level images. The first 10 images in each class of each training face dataset are considered for the training set. PCA is applied for the dimensionality reduction and for face space formation.

Table 8.1 provides the results, where the COIL-100 database is used as the test set and the training datasets are the face datasets (such as AR, FERET, ORL and Yale). It is evident from the table that no image in the COIL-100 dataset is classified under the boundary of the face space, i.e, 100% accurate result has been obtained.

### 8.5.2     Comparison between threshold selections by ROC-based and set estimation-based techniques

The method of threshold selection using set estimation is compared with the conventional ROC-based threshold selection for face authentication systems. Both the authentication methods are applied on the face space

**TABLE 8.1**: Face/non-face classification: COIL-100 as test set

| Training face dataset | Number of projected faces | Number of images inside face space | Number of images outside face space |
|---|---|---|---|
| AR | 450 | 0 | 100 |
| FERET | 550 | 0 | 100 |
| ORL | 400 | 0 | 100 |
| Yale | 100 | 0 | 100 |

formed by using the dimensionality reduction techniques. To record the comparative performance of the two methods, several dataset configurations are designed.

As a convention, the intersection point of FRR and FAR curves, designated as equal error rate (EER), is chosen as the threshold for the whole system. With the help of the threshold, the system decides either (i) accept the test image as face image or (ii) reject the test image as a non-face image or an imposter image. Since the ROC-based method has only one threshold for the system, for fair comparison between the ROC-based scheme and the proposed method, the global threshold of the scheme is considered.

### 8.5.2.1 Formation of training–validation–test set

Several combinations of training, validation and test (TVT) sets are executed. The COIL-100 dataset is added as validation and as test set in four TVT configurations. The configurations follow the division ratio of the Lausanne protocol, according to which the face datasets are divided into three parts, namely, (i) training set, (ii) evaluation set and (iii) test set. The training set is used to build the client model. The evaluation set is used to compute client and impostor scores designated by FAR and FRR. The test set consists of face images and non face images.

AR and ORL datasets are divided according to even and odd classes as AR1, AR2 and ORL1, ORL2, respectively. ORL1 and AR1 are the odd numbered classes (i.e., $1, 3, 5, 7, ...$) of face images whereas ORL2 and AR2 are the even numbered classes $(2, 4, 6, )$ of face images of the corresponding datasets. To construct the validation and test set for the non-face dataset, the COIL-100 dataset is divided into two sets namely, COIL1 (with first 50 classes) and COIL2 (with other 50 classes). First 10 images of the considered classes are taken for the experiments.

If number of correctly detected face images is defined as #cf and the number of correctly detected non-face images as #ncf and the total number of face and non-face images in the test dataset as #tf, then the detection rate for each such TVT combination is given by

$$detection\ rate = \frac{\#cf + \#cnf}{\#tf} \qquad (8.3)$$

**TABLE 8.2**: Training-validation-test set configurations

| TVT | Traing Set No. of faces | Evaluation set | | Test set | |
|---|---|---|---|---|---|
| | | Imposters | Clients | Imposters | Clients |
| TVT 1 | ORL 40 × 5 | Yale 15×6 | ORL 40 × 3 | Yale 15×5 | ORL 40 × 2 |
| TVT 2 | ORL 40 × 5 | AR 68 × 5 | ORL 40 × 3 | AR 68 × 5 | ORL 40 × 2 |
| TVT 3 | Yale 15 × 5 | ORL 40 × 5 | Yale 15× 3 | ORL 40 × 5 | Yale 15 × 3 |
| TVT 4 | AR1 34 × 5 | AR2 34×5 | AR1 34 × 3 | AR2 34×5 | AR1 34 × 2 |
| TVT 5 | ORL2 20 × 5 | ORL1 20 × 5 | ORL2 20 × 3 | ORL1 20 × 5 | ORL2 20 × 2 |
| TVT 6 | Yale 15 × 5 | COIL1 50 × 10 | Yale 15 × 3 | COIL2 50 × 10 | Yale 15 × 3 |
| TVT 7 | AR 68 × 5 | COIL1 50 × 10 | AR 68 × 3 | COIL2 50 × 10 | AR 68 × 2 |
| TVT 8 | ORL1 20 × 5 | COIL1 50 × 10 | ORL1 20 × 5 | COIL2 50 × 10 | ORL1 20 × 2 |
| TVT 9 | AR 68 × 5 | Yale 15×6 | AR 68 × 3 | COIL1 50 × 10 | AR 68 × 2 |
| TVT 10 | Yale 15 × 5 | AR 68 × 5 | Yale 15 × 3 | COIL2 50 × 10 | Yale 15 × 3 |

**TABLE 8.3**: Comparison table

| TVT | Set estimation method | | EER as threshold | |
|---|---|---|---|---|
| | Recognition rate | Global threshold | Recognition rate | Global threshold |
| 1 | 97.25 | 1.5437e+003 | 95 | 2.1987e+003 |
| 2 | 98 | 1.5437e+003 | 94 | 2.3567e+003 |
| 3 | 96 | 1.3012e+003 | 95 | 1.9997e+003 |
| 4 | 98 | 2.3044e+003 | 93 | 2.546e+003 |
| 5 | 98.25 | 1.4824e+003 | 96.25 | 1.8234e+003 |
| 6 | 97 | 1.3012e+003 | 94 | 2.5897e+003 |
| 7 | 96.5 | 1.2658e+003 | 95 | 1.6787e+003 |
| 8 | 97 | 1.8732e+003 | 94 | 2.1523e+003 |
| 9 | 98 | 1.9984e+003 | 96.5 | 2.1437e+003 |
| 10 | 98 | 1.3012e+003 | 95.5 | 1.3485e+003 |

In Table 8.3, the detection results show that the set estimation method outperforms the ROC-based threshold selection method for each TVT configuration. It may be noted from the table that the global thresholds of the ROC-based method are generally higher than the proposed scheme. Another significant observation is that the global threshold remains the same for each training dataset, however, in the ROC-based method the threshold varies as the number of impostors in the validation set changes. Note that the training sets in TVT1 and TVT2 are the same but the corresponding thresholds in the ROC-based method are different. A similar situation is observed in the case of TVT3, TVT6 and TVT10. It can also be noted that in case of TVT6 to TVT10 COIL dataset is used as the impostor set.

The global threshold for the ROC based scheme depends upon the number of members in the validation set (i.e., is number of clients and number of impostors) as well as on the type of the impostor set. The variation in the threshold obtained for different validation sets can be observed in Table 8.3 and also shown in Figure 8.3 for ORL dataset. In the ROC,in the referenced figure, the value of EER-based threshold (the cutting point of the FAR and FRR curves) varies, as the validation set changes from (a) to (b).

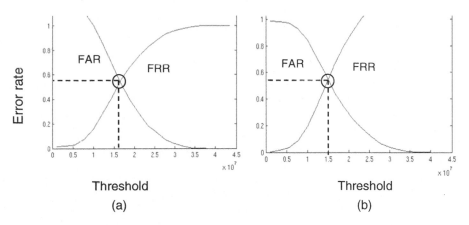

**FIGURE 8.3**: Sample ROCs for ORL dataset EER shifted as the validation set changes from (a) to (b)

The validity of the proposed method is established in the Tables 8.2 and 8.3, where the test images are objects or faces whose images are not present in the training database. Additionally, the more the number of non-face images considered for obtaining a ROC curve, the more is its complexity. Using the set estimation method, this drawback can be removed.

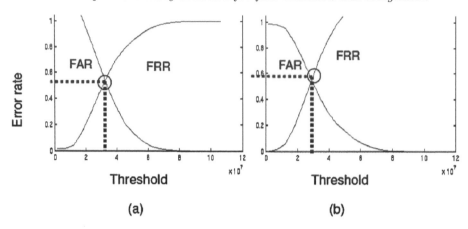

**FIGURE 8.4**: Sample ROCs for Yale dataset EER shifted as the validation set changes from (a) to (b)

## 8.6   Classification of face/non-face regions

The global threshold described can be applied on an image containing multiple faces with unconstrained background. The image may also contain non-face regions like dresses, hand portions and other background regions. The following algorithm can be applied for these types of images to classify the face and non-face portions for detection of the target face by classifying face and non-face regions of an image.

**Step 1**: PCA is performed on a standard face dataset to form the face space and also a threshold value at FAR = FRR.

**Step 2**: Let the number of faces in the given image be $I$ and the face regions be represented by $A_1, A_2, ......... A_p$. Let the number of non-face regions of the image having the same chosen size of the face portions be $q$ and they are represented by $B_1, B_2, ..........B_q$. The image portions (face and non-face regions) are chosen such in a way that (i) the size of the cropped face portions should be resized or transformed to the size of the standard dataset; (ii) there should not be any intersection among any two face regions. Two non-face regions may overlap, however, no non-face regions can have non empty intersection with a face region.

**Step 3**: Every region is projected in the preconstructed face space and classification between the face and non-face portions is achieved using the global threshold.

The process is tested on an FIA dataset containing multiple images of faces in a frame. The image under consideration is in Figure 8.5. The face images are of 64×64 in size. The cropped face portions of the given image are in Figure 8.6. Some of the cropped non-face portions of the given image are in Figure 8.7. The global threshold is obtained by the proposed algorithm. All face and non-face regions are classified with the help of the suggested procedure. Every non-face regions are found to be outside the face space.

**FIGURE 8.5**: Image containing multiple faces

**FIGURE 8.6**: Cropped face portions of the given image

**FIGURE 8.7**: Cropped non-face portions of the given image

The considered face portions all correspond to frontal views. An image containing the non face parts like hands, legs, necks is classified to be a non-face portions and depicts those image portions outside the face space. If we take any portion in which the face and non-face regions are mixed, the classification decision depends on the portion containing the face part. It has also been observed that the skin colors do not contribute much in the scheme.

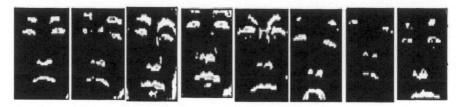

**FIGURE 8.8**: Face and non-face classification

---

## 8.7 Class specific thresholds of face-class boundaries for face recognition

To establish the usefulness of the set estimation method in the face classification problem, $M$ persons each having $N$ face images of the same size and same background are considered. The number of classes are the same as the number of persons. Initially any one of the feature extraction methods (say PCA) is applied to reduce the number of dimensions to $m$. Thus each face is now an $m$ dimensional sample point. For each one of $M$ classes, $N$ number of such $m$ dimensional vectors exist. Then for each class, MST of the respective $N$ vectors is calculated and its maximal edge weight is obtained. Let the maximal edge weight of the MST of the $i - th$ class be denoted by $\xi_i$. The recognition method for classifying a new point $x$ is given as

1. The total number of given vectors is $MN$. For each class $i$, find the minimum distance of $x$ with all the $N$ points in the class. Let the minimal distance be $\rho_i$.

2. If there exists an $i$ such that $\rho_i \leq \xi_i$, then put $x$ in the $ith$ class.

3. If $\rho_i > \xi_i$ for all $i$, then the given image does not fall in any one of the given face classes.

The process of applying local thresholds is depicted in Figure 8.1. In Figure 8.1, black discs denote the points in training set from class 1. Black discs with holes denote the points in training set from class 2. $\xi_1$ is the maximal edge weight of 4 edges of MST of 5 points in class 1. $\xi_2$ is the maximal edge weight of 4 edges of MST of 5 points in class 2. The point with rings is the point to be classified. Its nearest neighbour is at a distance $\rho_1$ from class 1, where $\rho_1 \leq \xi_1$. Note that the nearest neighbour of the point from class 2 is at a distance $\rho_2$ from it, where $\rho_2 > \rho_1$ and $\rho_2 > \xi_2$ . Thus, the point is classified to class 1.

## 8.8    Experimental design and result analysis

### 8.8.1    Description of face dataset

The proposed method has been used for face recognition and is tested over four well-known still face databases, namely, ORL [126], Yale [127], AR [128] and FERET [129] databases. Initially, to show the applications of the proposed method on gray level images, AR and FERET color image datasets are transformed to gray level image datasets. Later, in another set of experiments, AR and FERET datasets are considered, and the proposed method is extended to color images. Lastly, experiments are also carried out on video-face datasets.

For the purpose of reducing the dimensionality, feature extraction by using any one of the subspace methods is needed and all calculations are done in *face space*. The feature extraction methods explored here are principal component analysis (PCA), linear discriminant analysis (LDA) in combination with PCA, two dimensional PCA (2DPCA) and kernel PCA.

Several parameters are chosen for the experiments on face databases. These parameters are (i) number of training images, (ii) procedure for choosing those images and (iii) the number of reduced dimensions to be considered. All necessary information related to the construction of the face space with several face datasets, number of face classes used for training and number of reduced dimensions is described in Table 8.4.

**TABLE 8.4**: Formation of face space

| Database used | Number of classes | Number of images/class | Reduced Dimension of face space in % |
|---|---|---|---|
| FERET | 100 | 5 | 60 |
| AR | 68 | 5 | 60 |
| ORL | 40 | 5 | 60 |
| Yale | 15 | 5 | 60 |

After the dimensionality reduction, each face dataset is divided into two parts, namely the training dataset and the test dataset. A training dataset may contain points from every class of the dataset. In such a situation there is no imposter in the test data set, and therefore the experiment is for a closed test application. On the other hand, a training set may not contain points from every face class of the database resulting in a few impostors in the test set from the non-represented classes. This is an example of open test identification.

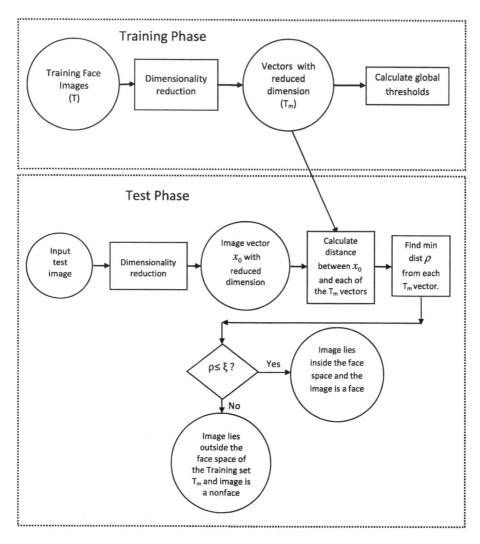

**FIGURE 8.9**: Diagram of training and testing phases of proposed algorithm

#### 8.8.1.1 Recognition rates

With respect to the application of set estimation theory in face recognition, the technique states that the term recognized should answer two different queries [130]. These are (1) whether the test sample belongs to the set of clients (open test case) and, (2) if yes, then which class member (or client) does the test sample represent.

Since the set estimation method derives the class-specific thresholds, the recognition system is in a position to accept the query image as a client or reject it as an imposter. For conducting an open test, the test set is divided into two sets namely, clients and impostors. The client set remains constant in the test set, while the numbers of impostors are gradually increased. If the number of correctly recognized client face images, denoted as (#cf), the number of correctly recognized impostors is (#ci), and the total number of clients and impostor images in the test dataset is (#tf), the recognition rate is defined as

$$recognition\ rate = (\frac{\#cf + \#ci}{\#tf}) \times 100 \qquad (8.4)$$

The method of threshold selection as discussed is applied on each training set of face points and the class-specific thresholds are computed for each class represented in the training set. The block diagram of the method of class-specific threshold based classification is given in Figure 8.9 for the training and testing phases.

### 8.8.2 Open test results considering imposters in the system

The proposed method does the classification at several levels and therefore the recognition rates may be defined accordingly. Four graphs corresponding to four different datasets are generated and are shown in Figure 8.10. Curves denoted by numerals indicate four subspace methods used, namely, PCA, KPCA, PCA-LDA, 2DPCA along with nearest neighbor thresholding. In the graphs, $y$ axis represents the recognition rates and $x$ axis depicts the number of attempts by the impostors. A set of curves in the Figure 8.10 (a)(v) also shows the recognition rates using the nearest neighbor (NN) classifier. No thresholding mechanism is used to obtain this rate. The curve (ii) denotes the variation in recognition rates using the class thresholds with PCA and LDA, curve (iii) indicates recognition rate using PCA and class threshold, curve ((iv) using KPCA and class threshold. In case of FERET data base shown in Figure 8.10(b), (i) indicates recognition rate using PCA and class threshold, (ii) for PCA LDA and class threshold, (iii) recognition rate with 2DPCA and class threshold, (iv) KPCA and class threshold, (v) PCA, LDA and nearest neighbor (NN) thresholding, and (vi) KPCA and NN.

The process starts with 40 impostors face images. As the number of impostor images is increased, the recognition rate drops in case of the neural network classifier. In contrast the performance of the set estimation method is better for each case. It is also established from the figures that the discussed classifier outperforms the non threshold-based systems. As the number of impostors increases, the recognition rates for other classifiers are seen to drop, whereas the recognition rate for the set estimation-based classifier is found to increase.

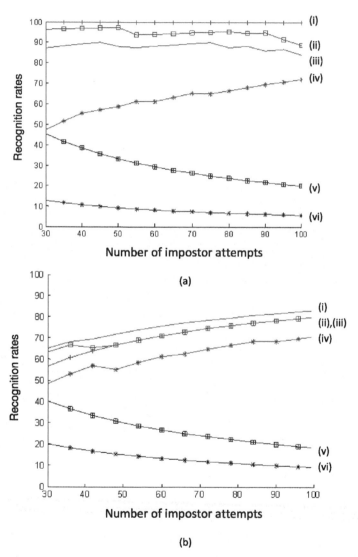

**FIGURE 8.10**: Open test results on (a)AR dataset (b) FERET dataset

### 8.8.3 Recognition rates considering only clients in the system

For the closed test, four subspace algorithms are considered to form the face spaces. Four datasets, FERET, AR, ORL and Yale are used in the experiment. The related results for recognition rates using the class specific threshold for classification, under different dimensionality reduction techniques are shown in Figure 8.11.

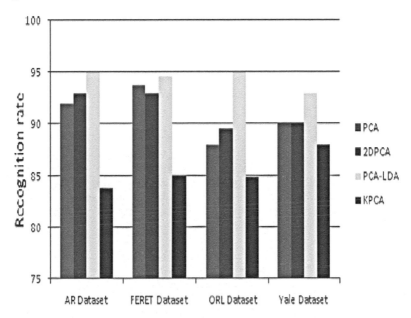

**FIGURE 8.11**: Closed test results on four different datasets on four different subspaces

---

MATLAB code for open and close face recognition

---

```
\texttt{function v2()
% finding recognition rate between test and
training face database
   clc
   clear all

   disp("*********** MENU ******************")
   disp("***** Choose your Dataset *********")
   disp("1. ARFACE 2.FERET 3. Yale 4. ORL 5.VITIMID")
   choice=input(" Enter your choice ");

   if choice==1
```

```
        load("arface.mat");
        noc=30;     % max number of class considered
        nocA=10;    % number of class in training set A
        ipc=13;          % images per class
        imseqA=1:2:13; % image sequence for training class
        imseqB=2:2:10; % image sequence for test class
        m=72;n=96; % mxn is the image size agter being scaled
elseif choice==2
        load("feret.mat");
        noc=30;     % max number of class considered
        nocA=10;    % number of class in training set A
        ipc=11;          % images per class
        imseqA=[5 6 7 8 9]; % image sequence for training class
        imseqB=[1 2 3 4 5]; % image sequence for test class
        m=64;n=43; % mxn is the image size agter being scaled
elseif choice==3
        load("yale.mat");
        noc=15;     % max number of class considered
        nocA=5;   % number of class in training set A
        ipc=11;          % images per class
        imseqA=1:2:11; % image sequence for training class
        imseqB=2:2:11; % image sequence for test class
        m=64;n=64; % mxn is the image size agter being scaled
elseif choice==4
        load("orl.mat");
        noc=40;     % max number of class considered
        nocA=10;    % number of class in training set A
        ipc=10;          % images per class
        imseqA=[1 3 5 7 9]; % image sequence for training class
        imseqB=[2 4 6 8 10]; % image sequence for test class
        m=56;n=46; % mxn is the image size agter being scaled
elseif choice==5
        load("vitimid.mat");
        noc=10;     % max number of class considered
        nocA=5;   % number of class in training set A
        ipc=40;          % images per class
        imseqA=1:2:40; % image sequence for training class
        imseqB=2:2:40; % image sequence for test class
        m=192;n=256; % mxn is the image size agter being scaled

else
        disp(" invlid choice ")
        return;
end


for pca_choice=1:4 % pca_choice==1 indicates PCA
                   % pca_choice==2 indicates 2D PCA
                   % pca_choice==3 indicates Kernel PCA
```

```
                       % pca_choice==4 indicates PCA_LDA
        plotCounter=0;
        nop=3;      % number of points within one mst edge

    for nocB=(nocA):noc
          %  number of class in test set B

          % step 0. preliminary calculations
          ipcA=length(imseqA);    %  images per class in  A
          ipcB=length(imseqB);    %  images per class in  B

            % step 1. read the training set A and test set B
            A=[];
            for classA=1:nocA
                for imgA=imseqA
                    col=(classA-1)*ipc+imgA;
                    A=[A set(:,col)];
                end
            end
            B=[];
            for classB=1:nocB
                for imgB=imseqB
                    col=(classB-1)*ipc+imgB;
                    B=[B set(:,col)];
                end
            end
            A=double(A);
            B=double(B);

            if pca_choice==1
% perform pca on train set A to find face space projection projA
                [ projA projB]=pca(A,B,size(A,2));
            elseif pca_choice==2  % 2D PCA
                % preprocessing
                clear projA;clear projB;

                for j=1:size(A,2)
                    train(:,:,j)=reshape(A(:,j),m,n);
                end
                for j=1:size(B,2)
                    test(:,:,j)=reshape(B(:,j),m,n);
                end
                % 2d pca
                [pA pB]=pca2d(train,test,n);

                % postprocessing
                for j=1:size(pA,3)
                    temp=pA(:,:,j);
                    projA(:,j)=temp(:);
```

```
                        end
                    for j=1:size(pB,3)
                        temp=pB(:,:,j);
                        projB(:,j)=temp(:);
                    end
                elseif pca_choice==3    % Kernel PCA
                    clear projA;clear projB;
                    [projA projB]=pcaKernel(A,B,size(A,2));
                elseif pca_choice==4    % PCA LDA
                    clear projA;clear projB;
                    [projA projB]=pcalda(A,B,noc,ipcA);
                end

                classA=1;
                for imgA=1:ipcA:size(projA,2)
                    exist=projA(:,imgA:imgA+ipcA-1);
                    % form the weighted graph wg, where
                    %weight=distance between nodes
                    wg=distMat(exist,exist);
                    [cost,next]=prim(wg,1);
% apply prim"s algo to get the MST

                    if nocB<=nocA
                    theta(classA)=max(cost);
% threshold for each class
                    else

                        theta(classA)=max(cost)/2;
                        % threshold for each class
                    end
                    classA=classA+1;
                end

            euDist=distMat(projA,projB);
        [correct wrong rate1]=clfr_nn(euDist,nocA,nocB);
  [correct wrong rate2]=clfr_minDist(euDist,nocA,nocB,theta);
  [correct wrong rate3]=clfr_maxHit(euDist,nocA,nocB,theta);

            plotCounter=plotCounter+1;
            nImpost=(nocB-nocA)*ipcB;
            xx(plotCounter)=nImpost;
            if pca_choice==1
                pca1(plotCounter)=rate1;
                pca2(plotCounter)=rate2;
                pca3(plotCounter)=rate3;
            elseif pca_choice==2
                pca2d1(plotCounter)=rate1;
                pca2d2(plotCounter)=rate2;
                pca2d3(plotCounter)=rate3;
```

```
        elseif pca_choice==3
            pcak1(plotCounter)=rate1;
            pcak2(plotCounter)=rate2;
            pcak3(plotCounter)=rate3;
        elseif pca_choice==4
            pcalda1(plotCounter)=rate1;
            pcalda2(plotCounter)=rate2;
            pcalda3(plotCounter)=rate3;

        end


    end % next nocB

  end % next pca_choice


  axis([20 100 0 100]);
  hold all;

  % plot for PCA
  plot(xx,pca1,"-b")  % classifier 1
  plot(xx,pca2,"-r")  % classifier 2


  % plot for 2D PCA
  plot(xx,pca2d1,"-b+")
  plot(xx,pca2d2,"-r+")

  % plot for Kernel PCA
   plot(xx,pcak1,"-b*")
   plot(xx,pcak2,"-r*")

  % plot for PCA LDA
  plot(xx,pcalda1,"-bs")
  plot(xx,pcalda2,"-rs")

 end



function euDist=distMat(A,B)
      for a=1:size(A,2)
          for b=1:size(B,2)
          % eucledian distance
              euDist(a,b)=sum((A(:,a)-B(:,b)).^2).^0.5;
          end
      end
 end
```

```
function [correct wrong rate]=clfr_nn(euDist,nocA,nocB)
      ipcA=floor(size(euDist,1)/nocA);
      ipcB=floor(size(euDist,2)/nocB);
      correct=0;
      wrong=0;
      for imgB=1:size(euDist,2)
            [val imgA]=min( euDist(:,imgB) );
            classB=floor((imgB-1)/ipcB)+1;
            classA=floor((imgA-1)/ipcA)+1;

            if(classA==classB)
                correct=correct+1;
            else
                wrong=wrong+1;
            end
      end
      rate=(correct*100)/(correct+wrong);
   end


function [correct wrong rate]=clfr_minDist(euDist,nocA,nocB,thresh)
      ipcA=floor(size(euDist,1)/nocA);
      ipcB=floor(size(euDist,2)/nocB);
      correct=0;
      wrong=0;
      for imgB=1:size(euDist,2)
            classB=floor((imgB-1)/ipcB)+1;
% find all candidate near images of imgB
            count=0;
            for imgA=1:size(euDist,1)
                  classA=floor((imgA-1)/ipcA)+1;
                  if euDist(imgA,imgB)<thresh(classA)
                      count=count+1;
                      candidateImg(count)=imgA;
                      candidateDist(count)=euDist(imgA,imgB);
                  end
            end
            if count==0 && classB>nocA % imposter
                correct=correct+1;
            elseif count==0 && classB<=nocA
                wrong=wrong+1;
            else
                [val ind]=min(candidateDist);
                 minImg=candidateImg(ind);
                 classA=floor((minImg-1)/ipcA)+1;

                if(classA==classB)
                      correct=correct+1;
                else
```

```
                                wrong=wrong+1;
                    end
                end
            end
            rate=(correct*100)/(correct+wrong);
        end



    function [correct wrong rate]=clfr_maxHit(euDist,nocA,nocB,thresh)
            ipcA=floor(size(euDist,1)/nocA);
            ipcB=floor(size(euDist,2)/nocB);
            correct=0;
            wrong=0;
            for imgB=1:size(euDist,2)
                classB=floor((imgB-1)/ipcB)+1;
                counter=zeros(1,nocA);
                for imgA=1:size(euDist,1)
                    classA=floor((imgA-1)/ipcA)+1;
                    if euDist(imgA,imgB)<thresh(classA)
                        counter(classA)=counter(classA)+1;
                    end

                end
                [val,classA]=max(counter);
% in which class imgB is classified max no of times
                if (classA==classB) || (val==0 && classB>nocA)
                    correct=correct+1;
                else
                        wrong=wrong+1;
                end
            end
            rate=(correct*100)/(correct+wrong);
        end
}
```

# Chapter 9

# Evolutionary design for face recognition

## 9.1   Introduction

Evolutionary pursuit (EP) [131] is a novel and adaptive representation method for image encoding and classification. EP seeks to learn an optimal basis for the dual purpose of data compression and pattern classification. The challenge for EP is to increase the generalization ability of the learning machine as a result of seeking the trade-off between minimizing the empirical risk encountered during training and narrowing the confidence interval for reducing the guaranteed risk during future testing on unseen images. EP implements strategies characteristic of genetic algorithms (GAs) for searching the space of possible solutions to determine the optimal basis. EP starts by projecting the original data into a lower dimensional whitened image space obtained from principal component analysis (PCA). Directed but random rotations of the basis vectors in this space are then searched by GAs where evolution is driven by a fitness function defined in terms of performance accuracy which is again termed as empirical risk for class separation.

Face recognition depends heavily on the particular choice of face features used by the classifier. One usually starts with a given set of features and then attempts to derive an optimal subset of features leading to high classification performance with the expectation that a similar performance will be displayed also in future trials on other datasets. The process of feature selection involves the derivation of salient features with the twin goals of reducing the amount of data used for classification and simultaneously providing enhanced discriminatory power. For optimal basis representation, most practical computational methods opt for both regression and classification by using parametrization in the form of a linear combination of basis functions. Since most practical methods use non-linear models, the determination of optimal kernels becomes a non-linear optimization problem. When the objective function lacks an analytical form suitable for gradient descent or the computation involved is prohibitively expensive, (directed) random search techniques for non-linear optimization and variable selection are used. These are similar to evolutionary computation and GAs. A brief introduction of genetic algorithms is provided in the next section.

## 9.2   Genetic algorithms

Genetic algorithms (GAs) are adaptive heuristic search algorithms based on the evolutionary ideas of natural selection and genetics. As such they represent an intelligent exploitation of a random search used to solve optimization problems. Although randomized, GAs are by no means random; instead they exploit historical information to direct the search into the region of better performance within the search space. The basic techniques of the GAs are designed to simulate the natural processes of survival of the fittest. GAs simulate the survival of the fittest among individuals over consecutive generations for solving a problem. Each generation consists of a population of character strings that are analogous to the chromosome. Each individual represents a point in a search space and a possible solution. The individuals in the population are then made to go through a process of evolution.

GAs are based on an analogy with the genetic structure and behaviour of chromosomes within a population of individuals using the the foundations that (a) individuals in a population compete for resources and mates, (b) those individuals most successful in each competition will produce more offspring than those individuals that perform poorly and (c) genes from so-called good individuals propagate throughout the population so that two good parents sometimes produce offspring that are better than either parent. Thus each successive generation becomes more suited to their environment.

In GAs, a population of individuals is maintained within a search space for a GA, each representing a possible solution to a given problem. Each individual is coded as a finite length vector of components, or variables, in terms of some alphabet, usually the binary alphabet (0,1). To continue the genetic analogy these individuals are likened to chromosomes and the variables are analogous to genes. Thus, equivalently, a solution is composed of several variables. A fitness score is assigned to each solution representing the abilities of an individual to compete. The individual with the optimal (or generally near optimal) fitness score is sought. The GA maintains a population of $n$ solutions with associated fitness values. New generations of solutions are produced containing, on average, more good genes than a typical solution in a previous generation. Each successive generation will contain better partial solutions than previous generations. Eventually, once the population has converged and is not producing offspring noticeably different from those in previous generations, the algorithm itself is said to have converged to a set of solutions to the problem.

## 9.2.1  Implementation

After an initial population is randomly generated, the algorithm evolves through three operators:

- **selection** which equates to survival of the fittest;

  - Give preference to better individuals, allowing them to pass on their genes to the next generation
  - The goodness of each individual depends on its fitness
  - Fitness may be determined by an objective function or by a subjective judgement

- **crossover** which represents mating between individuals

  - Prime distinguished factor of GA from other optimization techniques
  - Two individuals are chosen from the population using the selection operator. A crossover site along the bit strings is randomly chosen
  - The values of the two strings are exchanged up to this point. If $S1 = 000000$ and $S2 = 111111$ and the crossover point is 2 then $S1' = 110000$ and $S2' = 001111$
  - The two new offspring created from this mating are put into the next generation of the population
  - By recombining portions of good individuals, this process is likely to create even better individuals

- **mutation** which introduces random modifications

  - With some low probability, a portion of the new individuals will have some of their bits flipped
  - Its purpose is to maintain diversity within the population and inhibit premature convergence
  - Mutation alone induces a random walk through the search space
  - Mutation and selection (without crossover) create parallel, noise-tolerant, hill-climbing algorithms

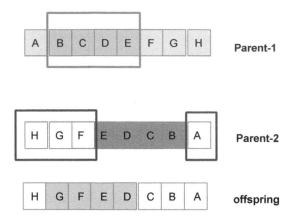

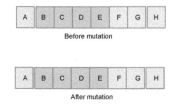

**FIGURE 9.1**: Parents and offspring generation during crossover

**FIGURE 9.2**: Mutation process

## 9.2.2 Algorithm

When the GA is implemented, it is usually done in a manner that involves the following cycle: Evaluate the fitness of all of the individuals in the population. Create a new population by performing operations such as crossover, fitness-proportionate reproduction and mutation on the individuals whose fitness has just been measured. Discard the old population and iterate using the new population. One iteration of this loop is referred to as a generation. The first generation (generation 0) of this process operates on a population of randomly generated individuals. From there on, the genetic operations, in concert with the fitness measure, operate to improve the population.

Algorithm GA is
// *start with an initial time*
$t := 0;$

// *initialize a usually random population of individuals*
initpopulation $P(t);$

// *evaluate fitness of all initial individuals of population*
evaluate $P(t)$;

// *test for termination criterion (time, fitness, etc.)*
while not done do

// *increase the time counter*
$t := t + 1$;

// *select a sub-population for offspring production*
$P' :=$ selectparents $P(t)$;

// *recombine the genes of selected parents*
recombine $P'(t)$;

// *perturb the mated population stochastically*
mutate $P'(t)$;

// *evaluate its new fitness*
evaluate $P'(t)$;

// *select the survivors from actual fitness*
$P :=$ survive $P, P'(t)$;

end GA.

---

## 9.3   Representation and discrimination

Efficient coding schemes for face recognition require both low-dimensional feature representations and enhanced discrimination abilities. The evolutionary method controls the reduction of both dimensionality and enhancement of discriminant power. Figure 9.3 illustrates a face recognition procedure using the evolutionary method.

### 9.3.1   Whitening and rotation transformation

After dimensionality reduction using PCA, the lower dimensional feature set $Z \in \Re^{m \times n}$ is derived as

$$Z = [Y_1 Y_2 \cdots Y_n] \tag{9.1}$$

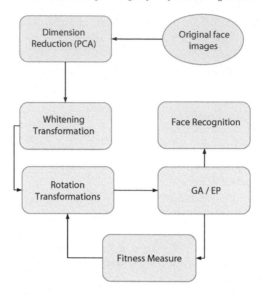

**FIGURE 9.3**: Block diagram illustrates the face recognition approach

where $n$ is the total number of training sets and $Y_i$s are the lower dimensional representation of the original dataset obtained by projecting the data on principal components.

A new feature set is evaluated by whitening transformation with $Z$ as

$$Z_n = \Lambda Z \tag{9.2}$$

where $\Lambda = diag\{\lambda_1^{-1/2}, \lambda_2^{-1/2} \cdots \lambda_n^{-1/2}\}$.

The rotation transformations are carried out in the whitened $m$ dimensional space in which the feature set $Z_n$ exists. Let $\Omega = [\epsilon_1 \epsilon_2 \cdots \epsilon_m]$ be the basis of this space where $\epsilon_1, \epsilon_2, \cdots, \epsilon_m$ are the unit vectors and $\Omega \in \Re^{m \times m}$. The evolutionary method searches for a reduced subset of some basis vectors rotated from $\epsilon_1, \epsilon_2, \cdots, \epsilon_m$ in terms of best discrimination and generalization performance. The rotation procedure is carried out by pairwise axes rotations. If the basis vectors $\epsilon_i$ and $\epsilon_j$ are rotated by $\alpha_k$ then a new basis $\zeta_1, \zeta_2, \cdots, \zeta_m$ is derived as

$$[\zeta_1 \zeta_2 \cdots \zeta_m] = [\epsilon_1 \epsilon_2 \cdots \epsilon_m] Q_k \tag{9.3}$$

where $Q_k \in \Re^{m \times m}$ is a rotation matrix.

There are total of $M = m(m-1)/2$ rotation angles correspond to $M$ pairs of basis vectors to be rotated. For the purpose of evolving an optimal basis for face recognition, it makes no difference if the angles are confined to $(0, \pi/2)$. The overall rotation matrix $Q \in \Re^{m \times m}$ is defined as

$$Q = Q_1 Q_2 \cdots Q_{m(m-1)/2} \tag{9.4}$$

The task of EP is to search for a face basis through the rotated axes defined in a properly whitened reduced dimensional space. Evolution is driven by a fitness function defined in terms of performance accuracy and class separation (scatter index). Accuracy indicates the extent to which learning has been successful so far, while the scatter index gives an indication of the expected fitness on future trials. Together, the accuracy and the scatter index give an indication of the overall performance ability. In analogy to the statistical learning theory, the scatter index is the conceptual analog for the capacity of the classifier and its use is to prevent overfitting. By combining these two terms together (with proper weights), GA can evolve balanced results and yield good recognition performance and generalization abilities.

One should also point out that just using more principal components (PCs) does not necessarily lead to better performance, since some PCs might capture the within-class scatter which is unwanted for the purpose of recognition. A search of the 20 and 30 dimensional whitened PCA spaces corresponding to the leading eigenvalues is logical, since it is in those spaces that most of the variations characteristic of human faces occur.

### 9.3.2  Chromosome representation and genetic operators

Different basis vectors are derived corresponding to different sets of rotation angles. GAs are used to search among the different rotation transformations and different combinations of basis vectors in order to find out the optimal subset of vectors (face basis), where optimality is defined with respect to classification accuracy and generalization ability. The optimal basis is evolved from a larger vector set $\{\zeta_1, \zeta_2, ..., \zeta_m\}$ rotated from a basis $\epsilon_1, \epsilon_2, ...\epsilon_m$ in $m$ dimensional space by a set of rotation angles $\alpha_1, \alpha_2, ....\alpha_{m(m-1)/2}$ with each angle in the range of $(0, \pi/2)$. If the angles are discretized with small enough steps, then GAs can be used to search this discretized space. GAs require the solutions to be represented in the form of bit strings or chromosomes.

### 9.3.3  The fitness function

Fitness values guide GAs on how to choose offspring for the next generation from the current parent generation. If $F = \alpha_1, \alpha_2, ....\alpha_{m(m-1)/2}; a_1, a_2, ..., a_m$ represents the parameters to be evolved by GA, then the fitness function is defined as

$$\xi(F) = \xi_a(F) + \lambda \xi_s(F) \tag{9.5}$$

where, $\xi_a(F)$ is the performance accuracy term, $\xi_s(F)$ is the class separation term and $\lambda$ is a positive constant that determines the importance of the second term relative to the first one.

$\xi_a(F)$ can be set at the number of faces correctly recognized as the top choice after rotation and selection of a subset of axes. $\xi_s(F)$ is the scatter measurement among the different classes. $\lambda$ is empirically chosen such that

$\xi_a(F)$ contributes more to the fitness than $\xi_s(F)$ does. Contributions of the two terms $\xi_a(F)$ and $\xi_s(F)$, however, work in the opposite directions on the fitness function. The performance accuracy term $\xi_a(F)$ tends to choose basis vectors which lead to small scatter, while the class separation term $\xi_s(F)$ favors basis vectors which cause large scatter. By combining those two terms together with proper $\lambda$, GA can evolve balanced results displaying good performance during both training and test trials. The rotation angles are set as $\alpha_1^k, \alpha_2^k, ..., \alpha_{m(m-1)/2}^k$ and the basis vectors after the transformation are $\xi_1^k, \xi_2^k, ..., \xi_m^k$. GA chooses $l$ vectors $\eta_1, \eta_2, ..., \eta_l$ from $\xi_1^k, \xi_2^k, ..., \xi_m^k$; then a new feature set $W \in \Re^{l \times n}$ is given by

$$W = [\eta_1 \eta_2 ... \eta_l]^T V \tag{9.6}$$

where $V$ is the whitened feature set.

Let $\omega_1, \omega_2, ..., \omega_L$ and $N_1, N_2, ..., N_L$ denote classes of images within each class, respectively. Let $M_1, M_2, ..., M_L$ and $M_0$ be the means of corresponding classes and global mean in the new feature space, $span[\eta_1, \eta_2, ..., \eta_l]$. It can now be estimated as

$$M_i = \frac{1}{N_1} \sum_{j=1}^{N_i} W_j^i, \quad i = 1, 2, ..., L \tag{9.7}$$

where $W_j^i, j = 1, 2, ..., N_i$ represents the sample images from class $\omega_i$ and

$$M_0 = \frac{1}{n} \sum_{i=1}^{L} N_i M_i \tag{9.8}$$

where $n$ is the total number of images for all the classes.

Hence $\zeta_s(F)$ can be calculated as

$$\zeta_s(F) = \sqrt{\sum_{i=1}^{L}(M_i - M_0)^T(M_i - M_0)} \tag{9.9}$$

GA provides the optimal solution $F^o = \alpha_1^o, \alpha_2^o, ..., \alpha_{m(m-1)/2}^o$. $Q$ represents the particular basis set corresponding to the rotation angles $\alpha_1^o, \alpha_2^o, ..., \alpha_{m(m-1)/2}^o$ and the column vectors of $Q$ are $\Theta_1, \Theta_2, ..., \Theta_m$. If $\Theta_{i1}, \Theta_{i2}, ..., \Theta_{il}$ are the basis vectors corresponding to $a_1^o, a_2^o, ..., a_m^o$, then the optimal basis $T \in \Re^{m \times l}$ can be expressed as

$$T = [\Theta_{i1} \Theta_{i2} ... \Theta_{il}] \tag{9.10}$$

where $i_j \in \{1, 2, ..., m\}$, $i_j \neq 1_k$ for $j \neq k$ and $l \leq m$.

### 9.3.4 The evolutionary pursuit algorithm for face recognition

The EP algorithm works as follows [131],

- Compute the eigenvector and eigenvalue matrices $\phi$ and $\Lambda$ of the covariance matrix $\Sigma_x$. Choose then the first $m$ leading eigenvectors from $\phi$ as basis vectors and project the original image set onto those vectors to form the feature set $Z$ in this reduced PCA space.

- Whiten the feature set $Z$ and the new feature set $V$ is derived.

- Set $[\epsilon_1...\epsilon_m]$ to be a $m \times m$ unit matrix $[\epsilon_1...\epsilon_m] = I_m$.

- Begin the evolution loop until the stopping criterion is met such that the fitness does not change further or the maximum number of trials is reached.

  - Sweep the $m(m-1)/2$ pairs of axes according to a fixed order to get the rotation angle set $\alpha_1^k, \alpha_2^k, ..., \alpha_{m(m-1)/2}^k$ from the individual chromosome representation and rotate the unit basis vectors $\epsilon_1, \epsilon_2, ..., \epsilon_m$ in this $m$ dimensional space to derive the new projection axes $\xi_1^k, ..., \xi_m^k$.

  - Compute the fitness value in the feature space defined by the $l$ projection axes $\eta_1, ...\eta_l$ which are chosen from the rotated set of basis vectors.

  - Find the sets of angles and the subsets of projection axes that maximize the fitness value, and keep those chromosomes as the best solutions so far.

  - Change the values of rotation angles and the subsets of the projection axes according to the GAs' genetic operators, and repeat the evolution loop.

- Carry out recognition using the face basis $T = [\Theta_{i1}, ..., \Theta_{il}]$.

With the face basis, $T = [\Theta_{i1}, ..., \Theta_{il}]$ the new feature set is derived as,

$$U = [U_1 U_2...U_n] = T^T V \qquad (9.11)$$

where $V$ is the whitened feature set and $U_i \in \Re^l$ is the feature vector corresponding to the $i$th face image.

The classification rule is specified as

$$\|U_i - U_k^o\|_2 = min\|U_i - U_j^o\|_2, \qquad U_i \in \omega_k \qquad (9.12)$$

The new face image $U_i$ is classified to class $\omega_k$ from which the Euclidean distance is minimum.

# Chapter 10

## Frequency domain correlation filters in face recognition

## 10.1   Introduction

While there have been varying and significant levels of performance achieved through the use of spatial 2D image data, the use of a frequency domain representation sometimes achieves better performance for the face recognition tasks. The use of the Fourier transforms allow to quickly and easily obtain raw frequency data which are significantly more discriminating (after appropriate data manipulation) than the raw spatial data, from which it is derived. One can further increase the discrimination ability through additional and specific feature extraction algorithms intended for use in the frequency domain. In the majority of cases, correlation filters [132] are used to achieve desired performances due to several advantages, such as 1) it has built-in shift invariance, 2) correlation filters are based on integration operation and thus offer graceful degradation of any impairment to the test face image, 3) correlation filters can be designed to exhibit attributes such as noise tolerance and high ability for discrimination and 4) finally design of correlation filter is derived from closed form expressions and thus physically realizable.

Correlation is a robust and general technique for pattern recognition. Ever since the first use of the optical correlator for implementing matched spatial filters by VanderLugt [133], researchers have been trying to develop better filters for the recognition of shapes, objects and faces. Such filters are popularly referred to as correlation filters [132] since they are designed for implementation in frequency plane correlators. Frequency domain face

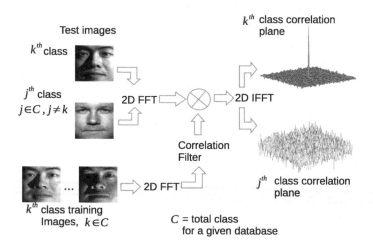

**FIGURE 10.1**: Basic frequency domain correlation technique for face recognition

recognition techniques are executed by cross correlating the Fourier transform of test face image with a synthesized template or filter, generated from the Fourier transform of training face images, as shown in Figure 10.1. The processing results in a correlation output via an inverse Fourier transform. An ideal correlation filter for face recognition would yield a sharp correlation peak for a perfect match of the correlation filter with a test face image present in the database. Such a test face is generally labelled as an authentic face. On the other hand if no such peak is found in the correlation plane the corresponding face images are labelled as impostors. Figure 10.2 shows the nature of a typical correlation plane in response to authentic and impostor face images. Generally a sharp peak is found in the case of authentic and no such peak is found in the case of an impostor.

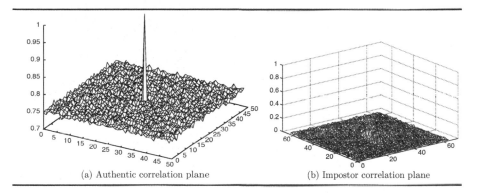

(a) Authentic correlation plane          (b) Impostor correlation plane

**FIGURE 10.2**: Correlation planes are shown for authentic and impostor face images.

The correlation output is searched for peak, and the relative height of this peak is analyzed to determine whether the test face is recognized or not. Figure 10.1 describes pictorially how the frequency domain correlation technique is carried out for face recognition using a correlation filter. As shown in Figure 10.1 the information of $N$ number of training images from $k$th face class ($k \in C$), out of total $C$ number of face classes for a given database, is Fourier transformed to form the design input for a $k$th correlation filter. The authentication of a test face is generally measured by a metric, called peak-to-sidelobe ratio (PSR) [134], which is measured from the correlation plane. In an ideal case a correlation peak with high value of PSR is obtained, when any Fourier transformed test face image of $k$th class is correlated with a $k$th correlation filter.

## 10.1.1 PSR calculation

A rectangular region (say $20 \times 20$ pixels) centered at the peak is extracted and used to compute PSR. A $5 \times 5$ rectangular region centered at the peak is masked out and the remaining annular region, shown in Figure 10.3, defined as the sidelobe region, is used to compute the mean and standard deviation of the sidelobes. The PSR is then calculated as

$$\text{PSR} = \frac{\text{peak-mean}}{\text{standard deviation}} \qquad (10.1)$$

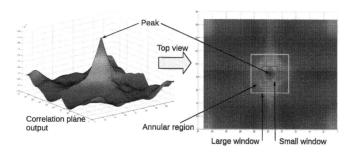

**FIGURE 10.3**: Pictorial representation of PSR metric evaluation from correlation plane output

The major correlation peak is below the threshold PSR value, if the test face image belongs to another class i.e. say, $j$th class where $j \in C$ and $j \neq k$. Evidently, the performance of the system in terms of recognition rate depends on the design of the correlation filter. The process given in Figure 10.1 can be mathematically summarized. Let $\mathbf{X}$ and $\mathbf{H}$ denote the 2D discrete Fourier transforms (DFTs) of 2D image $X$ and 2D filter $H$ in a spatial domain, respectively, and let $\mathbf{X}_i$ be the $i$th Fourier transformed test image of dimension $d_1 \times d_2$. The correlation output $G_i$ in the space domain in response to an $i$th image for the filter $\mathbf{H}$ can then be expressed as the inverse 2D DFT of frequency domain conjugate product as

$$G_i = FFT^{-1}[\mathbf{X}_i \circ \mathbf{H}^*], \qquad G_i \in \Re^{d_1 \times d_2} \qquad (10.2)$$

where $\circ$ represents the element wise array multiplication, * stands for complex conjugate operation and fast Fourier transform (FFT) is an efficient algorithm to perform a discrete Fourier transform (DFT).

## 10.2 A brief review on correlation filters

Development of correlation filters can be broadly categorized into two different classes: 1) linear constrained correlation filters and 2) linear unconstrained correlation filters. Constrained correlation filters are designed by specifying the output of filters for each training image. For $N$ training images, this results in $N$ constraints, which are typically much less than the number of free parameters, called the dimensionality of the filter. For this reason, many of these designs optimize some filter performance criterion while satisfying $N$ constraints. A general form of a constrained linear filter $\mathbf{h}$ is given by

$$\mathbf{h} = \bar{\mathbf{Q}}^{-1}\mathbf{A}(\mathbf{A}^+\bar{\mathbf{Q}}^{-1}\mathbf{A})^{-1}u \qquad (10.3)$$

where $\mathbf{A}$ is a matrix whose $N$ columns are $N$ frequency-domain training images ($\mathbf{x}_i$s) in vector form, $\bar{\mathbf{Q}}$ is a diagonal matrix and $u$ is an $N \times 1$ vector for the specified correlation output values for each training image.

Special cases of $\bar{\mathbf{Q}}$ result in well-known filter designs. These cases are listed in Table 10.1.

**TABLE 10.1**: Different values of diagonal matrix $\bar{\mathbf{Q}}$ result in different constrained correlation filters

| Filter type | Value of $\bar{\mathbf{Q}}$ from Equation 10.3 |
|---|---|
| ECPSDF [135] | $\bar{\mathbf{Q}} = \bar{\mathbf{I}}$ (Identity matrix) |
| MVSDF [136] | $\bar{\mathbf{Q}} = \bar{\mathbf{O}}$ |
| MACE [137] | $\bar{\mathbf{Q}} = \bar{\mathbf{D}},\ \bar{\mathbf{D}} = \sum\limits_{i=1}^{N}\bar{\mathbf{D}}_i,\ \text{where,}\ \bar{\mathbf{D}}_i - \bar{\mathbf{X}}_i\bar{\mathbf{X}}_i^*$ |
| OTSDF [138] | $\bar{\mathbf{Q}} - \alpha\bar{\mathbf{O}} + \sqrt{1-\alpha^2}\bar{\mathbf{D}}$ |
| MINACE [139] | $\bar{\mathbf{Q}} = max(\alpha\bar{\mathbf{O}}, \sqrt{1-\alpha^2}\bar{\mathbf{D}}_1, \cdots, \sqrt{1-\alpha^2}\bar{\mathbf{D}}_N)$ |

In Table 10.1, if $\bar{\mathbf{Q}}$ is replaced by $\bar{\mathbf{I}}$ (identity matrix), the design equation reduces to the ECP-SDF filter. The drawback of the ECP-SDF is that it cannot tolerate significant input noise. To achieve robustness to noise, a minimum variance synthetic discriminant function (MVSDF) filter is introduced [136]. Design equation of MVSDF is obtained by replacing $\bar{\mathbf{Q}}$ by $\bar{\mathbf{O}}$ where $\bar{\mathbf{O}}$ is a diagonal matrix containing the power spectral density of the noise. Thus, MVSDF minimizes the correlation output noise variance (ONV) while satisfying the correlation peak amplitude constraints. MVSDF controls only one point in the correlation map like the ECPSDF. and the variance of the noise matrix must be known beforehand in order to design the filter. However, if the latter is known exactly, MVSDF is impractical because it requires inverting a large noise covariance matrix [140, 132].

The minimum average correlation energy (MACE) filter is an attempt to control the entire correlation plane, where reduced correlation function levels are reduced at all points except at the origin of the correlation plane and a sharp correlation peak is obtained [137]. It has been shown that the operation is equivalent to minimizing the energy of the correlation function while satisfying intensity constraints at the origin. A closed form solution of the MACE filter is obtained by replacing $\bar{\mathbf{Q}}$ with $\bar{\mathbf{D}}$, as shown in Table 10.1, where $\bar{\mathbf{D}}_i$ is the power spectrum of the $i$th training image, and $\bar{\mathbf{D}}$ contains the average training power spectrum, and in $\bar{\mathbf{X}}_i = diag\{\mathbf{x}_i\}$. However, the MACE filter often suffers from two major drawbacks. First, there is again no built-in immunity to noise. Second, the MACE filter is often excessively sensitive to intra-class variations. Nevertheless, this filter establishes the utility of the frequency domain design approach for pattern recognition.

The optimal trade-off synthetic discriminant function (OTSDF) filter [138] includes a trade-off parameter $\alpha$ that allows the user to emphasize low output noise variance (ONV) ($\alpha$ closer to 1) or low average correlation energy (ACE) ($\alpha$ closer to 0). Setting $\alpha = 1$ yields MVSDF having minimum ONV but this usually exhibits a broad correlation peak. In contrast, setting $\alpha = 0$ yields the MACE filter, which has minimum ACE and produces sharp peak. However, the MACE filter is highly sensitive to noise and distortion.

The minimum noise and correlation energy (MINACE) filter [139] achieves an alternative compromise between these two extremes by using an envelope equal to or greater than the noise in the power spectra of the training image at each frequency. It may be noted that the trade-off parameter $\alpha$ appearing in the MINACE formulation in Table 10.1 is not a part of the traditional MINACE filter design as reported in [139]; rather, the value of $\bar{\mathbf{O}}$ is varied directly, since the input noise level is typically unknown. This difference is merely semantic; in practice, the same effect is achieved by varying either $\bar{\mathbf{O}}$ or $\alpha$. In both OTSDF and the MINACE filter designs, a single parameter $\alpha$ simultaneously accomplishes both these goals, because both the input noise level and the trade-off can be effected by scaling $\bar{\mathbf{O}}$ relative to $\bar{\mathbf{D}}$.

Studies have shown that hard constraints on correlation values at the origin are not only unnecessary but can be counterproductive [141]. Relaxing or removing such constraints might lead to a larger filter solution space. Also, the matrix inversion in the constrained design may be ill-conditioned, when highly similar training images are included. For these reasons, several unconstrained linear filter designs have been proposed. These designs maximize some measure of the average output on true-class training images while minimizing other criteria such as ONV and ACE. The maximum average correlation height (MACH) filter [140] is one such design, which achieves distortion tolerance by maximizing the similarity of the shapes of true-class correlation outputs over the training images. This maximization is realized by minimizing a dissimilarity metric known as the average similarity measure (ASM) for true class images. The design equation of the MACH filter is given in Table 10.2 where $\bar{\mathbf{S}}$ represents the measure of ASM.

Replacing $\bar{\mathbf{S}}$ by $\bar{\mathbf{D}}$ results in the unconstrained MACE (UMACE) filter [140] solution, given in Table 10.2. The unconstrained OTSDF (UOTSDF) filter [142] is a similar design that minimizes a trade-off between true-class ACE and ONV (as in the OTSDF design). The optimal trade-off approach is introduced in [143] by relating correlation plane metrics which results in the

**TABLE 10.2**: Design equations of unconstrained linear filters

| Filter type | Filter **h** |
|---|---|
| MACH [140] | $\bar{\mathbf{S}}^{-1}\mathbf{m}$ |
| UMACE [140] | $\bar{\mathbf{D}}^{-1}\mathbf{m}$ |
| UOTSDF [142] | $\{\alpha\bar{\mathbf{O}} + \beta\bar{\mathbf{D}}\}^{-1}\mathbf{m}$ |
| OTMACH [143] | $\{\alpha\bar{\mathbf{O}} + \beta\bar{\mathbf{D}} + \gamma\bar{\mathbf{S}}\}^{-1}\mathbf{m}$ |
| EMACH [144] | Dominant eigenvector of |
| | $\{\alpha\bar{\mathbf{I}} + (1-\alpha^2)^{1/2}\bar{\mathbf{S}}^\beta\}^{-1}\bar{\mathbf{C}}^\beta$ |
| EEMACH [145] | Dominant eigenvector of |
| | $\{\alpha\bar{\mathbf{I}} + (1-\alpha^2)^{1/2}\bar{\mathbf{S}}^\beta\}^{-1}\hat{\bar{\mathbf{C}}}^\beta$ |

OTMACH filter as given in Table 10.2, where, $\alpha$, $\beta$ and $\gamma$ are the non-negative optimal trade-off (OT) parameters.

In addition to the OTMACH filter, different variations of MACH filters were proposed. In [144] an extended MACH (EMACH) filter design is addressed by reducing the dependence on the average training image **m**. A tunable parameter $\beta$ (given in Table 10.2 is used to control this reduction. Two new metrics are used in the design: (1) the all-image correlation height (AICH), which takes into account of the filter output on **m** as well as on individual training images, and (2) a modified average similarity measure (MASM), which measures the average dissimilarity to the optimal output shape. This optimal shape reduces the dependence on **m** as realized by new a AICH metric. EMACH filter design also includes an ONV criterion to help in maintaining noise tolerance. A trade-off parameter $\alpha$, given in Table 10.2, is used to control the relative importance of the ONV and MASM criteria, where higher values of $\alpha$ correspond to greater emphasis on ONV and vice versa. If the covariance matrix $\bar{\mathbf{C}}^\beta$ is approximated by only its dominant eigenvectors, the eigenvalues yield a new matrix $\hat{\bar{\mathbf{C}}}^\beta$. The resulting filter solution is referred to as the eigen-extended MACH (EEMACH)filter [145].

A linear correlation output is an array of scalar output values from a linear discriminant applied to the input image at every shift. While limited in capability by their linear nature, linear correlation filters have the important advantage of efficient frequency domain computation. In addition to linear correlation filters several nonlinear correlation filters were developed. Design equations of some of the filters are given in Table 10.3. Special cases of nonlinear discriminant functions have been proposed in which some attractive computational properties of linear filters are retained by specialized implementation schemes.

Nonlinear correlation filters are designed in two ways. One class of design, termed as quadratic correlation filters (QCFs), is obtained by solving for a quadratic discriminant function in $d$-dimensional space, where $d$ is the number of pixels in the image. This quadratic discriminant can then be efficiently implemented as a set of linear filters via eigen decomposition. Several methods were proposed for solving the diagonal matrix in QCF design such as 1) subspace quadratic synthetic discriminant functions (SSQSDFs) [146], 2)Rayleigh quotient quadratic correlation filters (RQQCFs) [147], 3) minimum variance quadratic synthetic discriminant functions (MVQSDFs) [146] and 4) QCFs based on the Fukunaga Koontz transform [147].

**TABLE 10.3**: Some advanced correlation filters

| Filter | Design equation |
|---|---|
| GMACH [148, 144] | $\mathbf{h} = \{\delta\boldsymbol{\Omega} + \alpha\bar{\mathbf{O}} + \beta\bar{\mathbf{D}} + \gamma\bar{\mathbf{S}}\}^{-1}\mathbf{m}$ <br> where $\boldsymbol{\Omega}$ is $d^2 \times N$ matrix with rank $N$. |
| WaveMACH [149] | $\mathbf{h} = \{\bar{\mathbf{S}}^{-1}\mathbf{m}\}|\mathbf{H}(u,v)|^2$ <br> where $\mathbf{H}(u,v)$ is Mexican hat filter. |
| Log-WaveMACH [150] | $\mathbf{h} = \{\bar{\mathbf{S}}^{-1}\mathbf{m}\}|\mathbf{H}(u,v)|^2$ <br> log-polar transformation of training images. |
| ARCF [151] | $\mathbf{h} = (\bar{\mathbf{D}} + \epsilon\bar{\mathbf{I}})^{-1}\bar{\mathbf{X}}[\bar{\mathbf{X}}^+(\bar{\mathbf{D}} + \epsilon\bar{\mathbf{I}})^{-1}\bar{\mathbf{X}}]u$ <br> where $\epsilon = 0$ indicates MACE filter and <br> $\epsilon = \infty$ represents SDF filter. |
| CMACE [152] | $\mathbf{h} = \mathbf{V}^{-1}\mathbf{A}\{\mathbf{A}^+\mathbf{V}^{-1}\mathbf{A}\}^{-1}u$ in feature space <br> where $\mathbf{V} = \frac{1}{N}\sum_{i=1}^{N}\mathbf{V}_i$, $\mathbf{V}_i \triangleq$ correntropy matrix. |
| ActionMACH [153] | 3D version of MACH filter. |
| ASEF [154] | $\mathbf{H} = \frac{\mathbf{G}_i}{\mathbf{X}_i}$, <br> where $\mathbf{G}_i = FFT\{exp\{\frac{(m-m_i)^2+(n-n_i)^2}{\sigma^2}\}\}$, <br> is transformed Gaussian function at target location <br> $(m_i, n_i)$ and $\sigma \triangleq$ standard deviation. |
| MOSSE [155] | $\mathbf{H} = \frac{\sum_i \mathbf{G}_i\mathbf{X}_i^*}{\sum_i \mathbf{X}_i\mathbf{X}_i^*}$, for single training image <br> MOSSE filter = ASEF filter |
| MMCF [156] | $\mathbf{h} = \{\lambda\bar{\mathbf{I}} + (1-\lambda)(\bar{\mathbf{D}} - \mathbf{A}\mathbf{A}^+)\}^{-1/2}\tilde{\mathbf{A}}a$ <br> where $\tilde{\mathbf{A}} = [\tilde{\mathbf{x}}_1, \tilde{\mathbf{x}}_2, \cdots, \tilde{\mathbf{x}}_N]$, <br> and $\tilde{\mathbf{x}} \triangleq \{\lambda\bar{\mathbf{I}} + (1-\lambda)(\bar{\mathbf{D}} - \mathbf{A}\mathbf{A}^+)\}^{-1/2}\mathbf{x}_i$ <br> and $a$ [156] is evaluated by <br> sequential minimum optimization [157] technique. |

In contrast, in the second type of design, termed polynomial correlation filters (PCFs), are sets of linear filters applied to multichannel input images, whose outputs are subsequently summed to form a single output. Two variants of PCF are proposed: 1) constrained PCF (CPCF) [158] where CPCF design minimizes a weighted sum of ONV and ACE analogous to the OTSDF design, using the trade-off parameter $\alpha$ in a similar manner and 2) unconstrained PCF (UPCF) [159], where UPCF design maximizes the average correlation height (ACH), while minimizing a weighted sum of ONV and ACE.

In addition to the above-mentioned correlation filters some other correlation filters are designed to meet the demands of a specific application. Table 10.3 includes design equations of generalized MACH (GMACH) filter, wavelet modified MACH (WaveMACH) filter, log-transformed WaveMACH (Log-WaveMACH) filter, Action MACH filter, average exact synthetic function (ASEF) filter, correntropy MACE (CMACE) filter, adaptive robust correlation filter (ARCF), minimum output sum of squared error (MOSSE) filter and maximum margin correlation filter (MMCF). Detailed discussions on each of them are beyond the scope of the present work.

---

## 10.3    Mathematical background of correlation filter

### 10.3.1    ECPSDF filter design

Traditionally in the design of ECPSDF-type correlation filters, linear constraints are imposed on the training images to yield a known value at specific locations in the correlation plane. The classical ECPSDF [135, 160] filter is designed as a two-class problem, where the correlation values at the origin is set to 1 (may be selected to other values for multi-class problem) for training images from one class, generally authentic or true class, and to 0 for training images from other class or false class. The hope is that the resulting filter will yield values close to 1 for all images from class-1 and close to 0 for all images from class-2, and thus we can tell which class the input belongs to by looking at the value at the origin.

The above idea works well if all of the images (including non-training images) are always centered and thus we look only at the correlation value at the origin. However, one of the main advantages of using a correlation filter is its shift-invariance so we need not require that the input image be centered. However, if the image is not centered, we will need to determine to which pixel location the controlled values (of 1 and 0) have moved. The shift invariant property of the correlation filter is shown in Figure 10.4.

SDF can be formulated by a single matrix-vector equation denoted as

$$\mathbf{A}^{+}\mathbf{h} = u \tag{10.4}$$

where $\mathbf{A} = [\mathbf{x}_1, \mathbf{x}_2, ..., \mathbf{x}_N]$ is a $d \times N$ matrix with $N$ training Fourier transformed vectors as its columns, and $u = [u_{(1)}, u_{(2)}, \ldots, u_{(N)}]^T$ is an $N \times 1$ vector containing the desired peak values at the origin of correlation plane for the desired class, and $d$ is the total number of pixels present in one image.

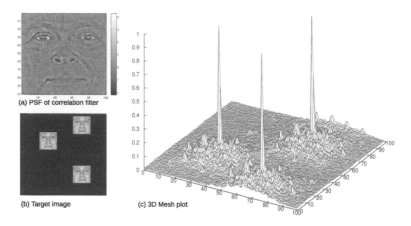

(a) PSF of correlation filter

(b) Target image

(c) 3D Mesh plot

**FIGURE 10.4**: Three distinct peaks are found at three different locations of the face images shown at left

**h** is the desired filter of size $d \times 1$ and the superscript $+$ indicates the complex conjugate transpose. However, since the number of training images $N$ is generally much smaller than the number of frequencies in the filters, the system of equations is under determined and many filters exist that satisfy the constraints in Equation 10.4. To find a unique solution, **h** is assumed to be a linear combination of the training images

$$\mathbf{h} = \mathbf{A}c \tag{10.5}$$

where the coefficient vector $c = [c_{(1)}, c_{(2)}, \cdots, c_{(N)}]^T$ of the linear combination is chosen to satisfy the deterministic constraints indicated in Equation 10.4.

The coefficient vector $c$ can be determined by substituting **h** in Equation 10.4 as

$$\mathbf{A}^+\mathbf{A}c = u \Rightarrow c = (\mathbf{A}^+\mathbf{A})^{-1}u \tag{10.6}$$

Substituting the solution for $c$ in Equation 10.5 leads to the SDF filter solution as

$$\mathbf{h}_{\mathrm{SDF}} = \mathbf{A}(\mathbf{A}^+\mathbf{A})^{-1}u \tag{10.7}$$

where $\mathbf{h}_{\mathrm{SDF}}$ is a $d \times 1$ filter vector in the transformed domain. Reshaping $\mathbf{h}_{\mathrm{SDF}}$ in proper row-column order yields $\mathbf{H}_{\mathrm{SDF}}$ of dimension $d_1 \times d_2$ in the transformed domain. Inverse Fourier transform of $\mathbf{H}_{\mathrm{SDF}}$ gives the solution for $H_{\mathrm{SDF}}$ in the space domain. The image of $H_{\mathrm{SDF}}$ is shown in Figure 10.5 where some representative training images are also shown with which the SDF filter is formulated. The composite nature of the filter is quite evident from Figure 10.5. Each of the training images is required to yield a value of 1.0 at the origin of the correlation plane.

The position of the pattern at the input is indicated by the location of

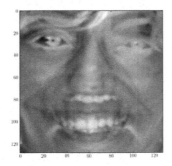

**FIGURE 10.5**: Image of SDF filter obtained from training samples from AR face database

the peak. The pattern is recognized when the peak value exceeds a certain threshold. However, the peak is surrounded by large sidelobes, which can lead to errors if they exceed the main peak. In fact, this occurs frequently in practice with a projection SDF filter since the filter design does not control any points in the correlation plane other than the origin. In practice, it is desirable to suppress the sidelobes to ensure a sharp and distinct correlation peak and reduce the chances of error. One way to achieve this is to minimize the energy in the correlation plane, which naturally includes the sidelobes. This leads to the further design of minimum average correlation energy filter or MACE filter.

### 10.3.2 MACE filter design

The MACE filter is designed to ensure sharp correlation peak and to allow easy detection in the full correlation plane as well as to control the correlation peak value. To achieve good detection, it is necessary to reduce the levels of correlation function at all points except at the origin of the correlation plane. Specifically, the value of the correlation function must have a user-specified value at the origin but the value is free to vary elsewhere. This is equivalent to minimizing the energy of the correlation function while satisfying intensity constraints at the origin. The correlation peak amplitude constraint for the MACE filter is the same as that considered in case of the SDF filter given in Equation 10.4. The correlation plane in response to $\mathbf{x}_i$ for the MACE filter $\mathbf{h}$ can be expressed in matrix-vector form as

$$\mathbf{g}_i = \bar{\mathbf{X}}_i^* \mathbf{h} \tag{10.8}$$

where $\bar{\mathbf{X}}_i$ represents a $d \times d$ diagonal matrix containing $i$th training vector $\mathbf{x}_i$ along its diagonal.

Hence the energy of the $i$th correlation plane can be formulated as

$$|\mathbf{g}_i|^2 = |\bar{\mathbf{X}}_i^* \mathbf{h}|^2 = \{\bar{\mathbf{X}}_i^* \mathbf{h}\}^+ \{\bar{\mathbf{X}}_i^* \mathbf{h}\} = \mathbf{h}^+ \bar{\mathbf{X}}_i \bar{\mathbf{X}}_i^* \mathbf{h} = \mathbf{h}^+ \bar{\mathbf{D}}_i \mathbf{h} \tag{10.9}$$

where $\bar{\mathbf{D}}_i = \bar{\mathbf{X}}_i\bar{\mathbf{X}}_i^*$ is a $d \times d$ diagonal matrix containing a power spectrum corresponding to $\mathbf{x}_i$.

For all $i = 1, 2, \cdots, N$ the average correlation energy (ACE) is given by,

$$\text{ACE} = \frac{1}{N}\sum_{i=1}^{N}|\mathbf{g}_i|^2 = \frac{1}{N}\sum_{i=1}^{N}\mathbf{h}^+\bar{\mathbf{D}}_i\mathbf{h} = \mathbf{h}^+\bar{\mathbf{D}}\mathbf{h} \qquad (10.10)$$

where $\bar{\mathbf{D}}$ represents a $d \times d$ diagonal matrix containing an average power spectrum along its diagonal and is given by

$$\bar{\mathbf{D}} = \frac{1}{N}\sum_{i=1}^{N}\bar{\mathbf{D}}_i = \frac{1}{N}\sum_{i=1}^{N}\bar{\mathbf{X}}_i\bar{\mathbf{X}}_i^* \qquad (10.11)$$

Therefore to synthesize the MACE filter, an attempt is made to minimize ACE given in Equation 10.10 while meeting the linear constraints in Equation 10.4. The minimum average correlation energy (MACE) filter [137] minimizes ACE in Equation 10.10 subject to the hard constraints in Equation 10.4. This is equivalent to a constrained quadratic optimization problem where the quadratic function $\mathbf{h}^+\bar{\mathbf{D}}\mathbf{h}$ is minimized subject to the linear conditions $\mathbf{A}^+\mathbf{h} = u$. This constrained quadratic optimization problem can be solved by using the method of Lagrange multipliers.

### 10.3.2.1    Constrained optimization with Lagrange multipliers

The method of Lagrange multipliers is useful for minimizing a quadratic function subject to a set of linear constraints. Suppose that $\mathbf{A} = [\mathbf{x}_1\mathbf{x}_2\cdots\mathbf{x}_N]$ is a $d \times N$ matrix with vectors $\mathbf{x}_i$ of length $d$ as its columns, and $u = [u_{(1)}, u_{(2)}, \ldots, u_{(N)}]^T$ is $N$ constants. We wish to determine the real vector $h$ which minimizes the quadratic term $\mathbf{h}^+\bar{\mathbf{D}}\mathbf{h}$ while satisfying the linear equations $\mathbf{A}^+\mathbf{h} = u$. Towards this end, we form the functional

$$\phi = \mathbf{h}^+\bar{\mathbf{D}}\mathbf{h} - 2\lambda_1(\mathbf{x}_1^T\mathbf{h} - u_{(1)}) - 2\lambda_2(\mathbf{x}_2^T\mathbf{h} - u_{(2)}) - \cdots 2\lambda_N(\mathbf{x}_N^T\mathbf{h} - u_{(N)}) \quad (10.12)$$

where the scalar parameters $\lambda_1, \lambda_2, \cdots \lambda_N$ are known as the Lagrange multipliers. These multipliers allow us to convert a constrained extremum problem into an unconstrained extremum problem. Setting the gradient of $\phi$ with respect to $\mathbf{h}$ to zero yields

$$2\bar{\mathbf{D}}\mathbf{h} - 2(\lambda_1\mathbf{x}_1 + \lambda_2\mathbf{x}_2 + \cdots, \lambda_N\mathbf{x}_N) = 0 \qquad (10.13)$$

Define $\lambda = [\lambda_1, \lambda_2, \cdots, \lambda_N]^T$; Equation 11.30 can be redrawn as

$$\bar{\mathbf{D}}\mathbf{h} - \mathbf{A}\lambda \Rightarrow \mathbf{h} = \bar{\mathbf{D}}^{-1}\mathbf{A}\lambda \qquad (10.14)$$

Substituting the value of $\mathbf{h}$ in constrained equation $\mathbf{A}^+\mathbf{h} = u$ yields

$$\mathbf{A}^+\bar{\mathbf{D}}^{-1}\mathbf{A}\lambda = u$$
$$or, \lambda = (\mathbf{A}^+\bar{\mathbf{D}}^{-1}\mathbf{A})^{-1}u \qquad (10.15)$$

Substituting the expression of $\lambda$ in Equation 10.14, the solution of the constrained optimization problem will be obtained as

$$\mathbf{h} = \bar{\mathbf{D}}^{-1}\mathbf{A}(\mathbf{A}^{+}\bar{\mathbf{D}}^{-1}\mathbf{A})^{-1}u \qquad (10.16)$$

Thus the optimum solution of the MACE filter is obtained as

$$\mathbf{h}_{\text{MACE}} = \bar{\mathbf{D}}^{-1}\mathbf{A}(\mathbf{A}^{+}\bar{\mathbf{D}}^{-1}\mathbf{A})^{-1}u \qquad (10.17)$$

The peak is very sharp with low sidelobes. MACE filters have been shown to be effective for finding training images in background and clutter, and they generally produce very sharp correlation peaks. The MACE filter is the first filter that attempted to control the entire correlation plane. However, there are two main drawbacks. First, there is no in-built immunity to noise. Second, the MACE filters are often excessively sensitive to intra-class variations. Nevertheless, the MACE filters paved the way for the frequency domain analysis and development of correlation filters, and set the stage for subsequent developments.

### 10.3.3 MVSDF filter design

MVSDF minimizes the correlation output noise variance (ONV) in $\mathbf{h}^{+}\bar{\mathbf{O}}\mathbf{h}$, where $\bar{\mathbf{O}}$ is the diagonal matrix whose diagonal entries are the noise power spectral density while satisfying the constraints of correlation peak amplitude as given in Equation 10.4. This is equivalent to optimizing a quadratic function subject to linear constraints. The method of Lagrange multipliers readily yields the following MVSDF filter solution as

$$\mathbf{h}_{\text{MVSDF}} = \bar{\mathbf{O}}^{-1}\mathbf{A}(\mathbf{A}^{+}\bar{\mathbf{O}}^{-1}\mathbf{A})^{-1}u \qquad (10.18)$$

The projection SDF filter is a special case obtained when the noise is white (i.e., $\bar{\mathbf{O}}$ is the identity matrix). Thus, the projection SDF filter is the optimum filter for recognizing the training images in the presence of additive white noise. The MACE filter yields sharp peaks that are easy to detect while the MVSDF filter is designed to provide robustness to noise. When there is only one training image, the MACE filter becomes the inverse filter, whereas the MVSDF filter becomes the matched filter. Since both attributes are important in practice, it is desirable to formulate a filter that possesses the ability to produce sharp peaks and behaves robustly in the presence of noise. This leads to the formulation of the optimal trade-off filter design.

### 10.3.4 Optimal trade-off (OTF) filter design

Due to the minimization criteria of ACE of the MACE filter, a sharp correlation peak is possible by suppressing the sidelobes as this filter emphasizes the high frequency components. However, the MACE filter can result in poor intra-class recognition of images which are not included in the training set. Moreover, the MACE filter is often excessively sensitive to noise as there is no in-built immunity to noise. To get a sharp correlation peak with suppressed noise, the MACE filter is combined with MVSDF. The technique resulted in the design of an optimal trade-off function (OTF) [138]. The optimum solution of OTF is given by

$$h_{OTF} = \bar{\mathbf{T}}^{-1}\mathbf{A}(\mathbf{A}^{+}\bar{\mathbf{T}}^{-1}\mathbf{A})^{-1}u \qquad (10.19)$$

where $\bar{\mathbf{T}} = \alpha\bar{\mathbf{D}} + \sqrt{1-\alpha^2}\bar{\mathbf{O}}$, $0 \leq \alpha \leq 1$. $\alpha$ is used as a controlling trade-off parameter, i.e., for $\alpha = 0$ leads to MVSDF and $\alpha = 1$ leads to the MACE filter.

## 10.3.5   Unconstrained correlation filter design

The SDF-type filters have assumed that the distortion tolerance of a filter could be controlled by explicitly specifying desired correlation peak values for training images. Unlike the SDF-type filter, in the case of designing unconstrained correlation filters no hard constraints (such as 1 for authentic and 0 for impostor) at the correlation planes are specified.

There are several observations that motivate this approach. First, non-training images always yield different values from those specified and achieved for the training images. Second, no formal relationship exists between the constraints imposed on the filter output and its ability to tolerate distortions. Distortion tolerance may also improve when the number of images in the training set is increased. This represents one method for reducing the sensitivity of the MACE filter. However, filter synthesis becomes computationally difficult, and a sufficiently large number of training images may not be available. There is no need for such a constraining assumption. In fact, once we realize that these pre-specified values are designated only for training images and not for test images, the justification for using this assumption decreases even more. Thus by removing the hard constraints, the number of possible solutions can be increased, thus improving the chances of finding a filter with better performance. In addition, the filters can be designed to offer good performance in the presence of noise and background clutter while maintaining relatively sharp correlation peaks for easy detection of the output. The above considerations lead to formulating further designs of unconstrained correlation filters.

### 10.3.5.1   MACH filter design

The unconstrained correlation filter offers improved distortion tolerance as during the design phase of such filters the training images are not treated as deterministic representations of the image, but as samples of a class whose characteristic parameters are used in encoding the filter [140]. In order to achieve this, an optimal shape of correlation plane $\mathbf{f}$ (in vector form of dimension $d \times 1$ ) is required and the deviation of the $i$th correlation plane in Equation 10.8 from the ideal shape vector $\mathbf{f}$ will be minimized. This deviation can be quantified in terms of average squared error (ASE) as

$$\text{ASE} = \frac{1}{N} \sum_{i=1}^{N} |\mathbf{g}_i - \mathbf{f}|^2 \tag{10.20}$$

Minimizing ASE by setting $\nabla_{\mathbf{f}}(ASE) = 0$ the optimum shape vector is obtained as

$$\mathbf{f}_{\text{opt}} = \frac{1}{N} \sum_{i=1}^{N} \mathbf{g}_i = \frac{1}{N} \sum_{i=1}^{N} \bar{\mathbf{X}}_i^* \mathbf{h} = \bar{\mathbf{M}}^* \mathbf{h} \tag{10.21}$$

where $\bar{\mathbf{M}} = \frac{1}{N} \sum_{i=1}^{N} \bar{\mathbf{X}}_i$. Equation 10.21 represents the average correlation plane and $\bar{\mathbf{M}}$ is the average training image expressed in diagonal form. The average correlation plane $\bar{\mathbf{M}}^* \mathbf{h}$ offers minimum ASE out of all possible reference shapes and hence least distortion in the squared error sense is achieved.

The average similarity measure (ASM) is a mean square error measure of distortions (variations) in the correlation surfaces relative to an average shape. In an ideal situation, all correlation surfaces produced by a distortion-invariant filter (in response to a valid input pattern) would be the same, and the ASM would be zero. In practice, minimizing the ASM improves the stability of the filter's output in response to distorted input images. It is now needed to formulate the ASM as the performance criterion for distortion invariant filter synthesis.

Mathematically ASM is obtained from Equation 10.20 by substituting $\mathbf{f} = \mathbf{f}_{\text{opt}} = \bar{\mathbf{M}}^* \mathbf{h}$ and $\mathbf{g}_i = \bar{\mathbf{X}}_i^* \mathbf{h}$ as

$$\begin{aligned}
\text{ASM} &= \frac{1}{N} \sum_{i=1}^{N} |\bar{\mathbf{X}}_i^* \mathbf{h} - \bar{\mathbf{M}}^* \mathbf{h}|^2 = \mathbf{h}^+ \bar{\mathbf{S}} \mathbf{h} \\
&= \frac{1}{N} \sum_{i=1}^{N} (\bar{\mathbf{X}}_i^* \mathbf{h} - \bar{\mathbf{M}}^* \mathbf{h})^+ (\bar{\mathbf{X}}_i^* \mathbf{h} - \bar{\mathbf{M}}^* \mathbf{h}) \\
&= \mathbf{h}^+ \left[ \frac{1}{N} \sum_{i=1}^{N} (\bar{\mathbf{X}}_i - \bar{\mathbf{M}})(\bar{\mathbf{X}}_i - \bar{\mathbf{M}})^* \right] \mathbf{h} \\
&= \mathbf{h}^+ \bar{\mathbf{S}} \mathbf{h} \tag{10.22}
\end{aligned}$$

where

$$\bar{\mathbf{S}} = \frac{1}{N} \sum_{i=1}^{N} (\bar{\mathbf{X}}_i - \bar{\mathbf{M}})(\bar{\mathbf{X}}_i - \bar{\mathbf{M}})^* \tag{10.23}$$

is a $d \times d$ diagonal matrix measuring the similarity of the training images to the class mean in the frequency domain.

In addition to being distortion-tolerant, a correlation filter must yield large peak values to facilitate detection of the pattern and to locate its position. Towards this end, we maximize the filter's average response to the training images. However, unlike the SDF filters, no hard constraints are imposed on the filter's response to training images at the origin. Rather, we simply desire that the filter should yield a high peak on average over the entire training set. This condition is met by maximizing the average correlation height (ACH) criterion defined as follows

$$\text{ACH} = \frac{1}{N} \sum_{i=1}^{N} \mathbf{x}_i^+ \mathbf{h} = \mathbf{m}^+ \mathbf{h} \tag{10.24}$$

where

$$\mathbf{m} = \frac{1}{N} \sum_{i=1}^{N} \mathbf{x}_i \tag{10.25}$$

represents the mean vector corresponding to training vectors $\mathbf{x}_i$ for all $i = 1, 2, \cdots, N$.

The peak intensity of the average correlation plane is written as

$$|\text{ACH}|^2 = |g(0,0)|^2 = |\mathbf{m}^+\mathbf{h}|^2 = \mathbf{h}^+\mathbf{m}\mathbf{m}^+\mathbf{h} \tag{10.26}$$

The behaviour of the average correlation plane is explicitly optimized by minimizing ASM and maximizing peak value. Hence the criterion to be optimized to improve distortion tolerance is given by

$$J(\mathbf{h}) = \frac{\mathbf{h}^+\mathbf{m}\mathbf{m}^+\mathbf{h}}{\mathbf{h}^+\bar{\mathbf{S}}\mathbf{h}} \tag{10.27}$$

where $J(\mathbf{h})$ is called the Rayleigh quotient.

The filter of interest $\mathbf{h}$ maximizes the average correlation height criteria and thus is called a maximum average correlation height (MACH) filter. The MACH filter maximizes the relative height of average correlation peak with respect to expected distortions. Since $J(\mathbf{h})$ in Equation 10.27 results in a small denominator, the filter $\mathbf{h}$ reduces the ASM given in Equation 10.22. The optimum filter is found by setting the gradient of $J(\mathbf{h})$ with respect to $\mathbf{h}$ to zero as follows

$$\begin{aligned} \nabla_h\{J(\mathbf{h})\} &= \frac{(\mathbf{h}^+\bar{\mathbf{S}}\mathbf{h})(2\mathbf{m}\mathbf{m}^+\mathbf{h}) - \mathbf{h}^+\mathbf{m}\mathbf{m}^+\mathbf{h}(2\bar{\mathbf{S}}\mathbf{h})}{(\mathbf{h}^+\bar{\mathbf{S}}\mathbf{h})^2} = 0 \\ &= \frac{\mathbf{m}\mathbf{m}^+\mathbf{h}}{\mathbf{h}^+\bar{\mathbf{S}}\mathbf{h}} - \frac{\mathbf{h}^+\mathbf{m}\mathbf{m}^+\mathbf{h}(\bar{\mathbf{S}}\mathbf{h})}{(\mathbf{h}^+\bar{\mathbf{S}}\mathbf{h})^2} = 0 \end{aligned} \tag{10.28}$$

This can be simplified to

$$\mathbf{m}\mathbf{m}^+\mathbf{h} = \lambda\bar{\mathbf{S}}\mathbf{h} \tag{10.29}$$

where

$$\lambda = \frac{\mathbf{h}^+\mathbf{m}\mathbf{m}^+\mathbf{h}}{\mathbf{h}^+\bar{\mathbf{S}}\mathbf{h}} \tag{10.30}$$

is the scalar identical to $J(\mathbf{h})$. Considering $\bar{\mathbf{S}}$ is invertible[1] Equation 10.29 can be rewritten as

$$\bar{\mathbf{S}}^{-1}\mathbf{mm}^{+}\mathbf{h} = \lambda\mathbf{h} \tag{10.31}$$

Now Equation 10.29 represents a generalized eigenvalue problem and from Equation 10.31 it can be stated that $\mathbf{h}$ is the eigenvector of $\bar{\mathbf{S}}^{-1}\mathbf{mm}^{+}$ with corresponding eigenvalue $\lambda$. Since $\lambda$ in Equation 10.30 is identical to $J(\mathbf{h})$ as shown in Equation 10.27 the eigenvector corresponding to the largest eigenvalue $\lambda$ is to be selected to maximize $J(\mathbf{h})$. Since $\mathbf{mm}^{+}$ is the outer product of a vector's $\bar{\mathbf{S}}^{-1}\mathbf{mm}^{+}$ has only one nonzero eigenvalue. The corresponding eigenvector is then the obvious choice for the optimum filter and can be found by substituting $\mathbf{m}^{+}\mathbf{h} = \mu$ (a scalar) in Equation 10.31 so that

$$\mu\bar{\mathbf{S}}^{-1}\mathbf{m} = \lambda\mathbf{h} \tag{10.32}$$

or

$$\mathbf{h}_{\text{MACH}} = \frac{\mu}{\lambda}\bar{\mathbf{S}}^{-1}\mathbf{m} \tag{10.33}$$

where $\mathbf{h}_{\text{MACH}}$ is the desired MACH filter, the transformed class-dependent mean image.

### 10.3.5.2   UMACE filter design

An interesting property of the MACH filter is that sharp peaks are obtained for true-class images even though their correlation energy is not explicitly minimized. The reason can be understood by expanding the ASM expression as

$$
\begin{aligned}
\text{ASM} &= \mathbf{h}^{+}\left[\frac{1}{N}\sum_{i=1}^{N}(\bar{\mathbf{X}}_{i} - \bar{\mathbf{M}})(\bar{\mathbf{X}}_{i} - \bar{\mathbf{M}})^{*}\right]\mathbf{h} \\
&= \mathbf{h}^{+}\left[\frac{1}{N}\sum_{i=1}^{N}\bar{\mathbf{X}}_{i}\bar{\mathbf{X}}_{i}^{*}\right]\mathbf{h} - \mathbf{h}^{+}\bar{\mathbf{M}}\bar{\mathbf{M}}^{*}\mathbf{h} \\
&= \mathbf{h}^{+}\bar{\mathbf{D}}\mathbf{h} - \mathbf{h}^{+}\bar{\mathbf{M}}\bar{\mathbf{M}}^{*}\mathbf{h} \\
&= \text{ACE} - \mathbf{h}^{+}\bar{\mathbf{M}}\bar{\mathbf{M}}^{*}\mathbf{h}
\end{aligned} \tag{10.34}
$$

Clearly, ASM includes the ACE term $\mathbf{h}^{+}\bar{\mathbf{D}}\mathbf{h}$, and therefore its minimization influences the correlation energies of the true-class images. In fact, the minimization of ASM can be viewed as a generalization of the MACE criterion. If $\mathbf{h}^{+}\bar{\mathbf{M}}\bar{\mathbf{M}}^{*}\mathbf{h}$ is small, then ASM $\approx$ ACE, and the performances of filters based on the two criteria are comparable.

The conventional MACE filter is also related to the MACH filter in a similar way. Replacing $\bar{\mathbf{S}}$ by $\bar{\mathbf{D}}$ in Equation 10.33, the filter expression becomes

$$\mathbf{h} = \bar{\mathbf{D}}^{-1}\mathbf{m} \tag{10.35}$$

---

[1]It is assumed that the training vectors are linearly independent to each other.

The MACE filter expression is

$$\mathbf{h}_{\mathrm{MACE}} = \bar{\mathbf{D}}^{-1}\mathbf{A}(\mathbf{A}^+\bar{\mathbf{D}}^{-1}\mathbf{A})^{-1}u = \bar{\mathbf{D}}^{-1}\mathbf{A}\mathbf{b} = \bar{\mathbf{D}}^{-1}\hat{\mathbf{x}} \qquad (10.36)$$

where $\hat{\mathbf{x}}$ is the weighted average image with weights $\mathbf{b} = (\mathbf{A}^+\bar{\mathbf{D}}^{-1}\mathbf{A})^{-1}u$ chosen to satisfy hard constraints on the training images.

As evident in Eqs.10.35 and 10.36, the MACH filter is of the same form as the MACE filter when the $\mathbf{h}^+\bar{\mathbf{M}}\bar{\mathbf{M}}^*\mathbf{h}$ term is dropped. Both filters minimize the same criterion (namely, ACE) although the latter does so under constraints. The special MACH filter obtained by dropping the $\mathbf{h}^+\bar{\mathbf{M}}\bar{\mathbf{M}}^*\mathbf{h}$ term is therefore referred to as the unconstrained MACE filter or the UMACE filter as

$$\mathbf{h}_{\mathrm{UMACE}} = \bar{\mathbf{D}}^{-1}\mathbf{m} \qquad (10.37)$$

### 10.3.5.3   OTMACH filter design

It has been shown in [143, 140] that the MACH filter and its other variants, most notably the optimal trade-off MACH (OTMACH) filter, are very powerful correlation filter algorithms. In practice, other performance measures like ACE and ONV are also considered to balance the system performance for different application scenarios. Optimal trade-off approach is introduced in [143] by relating correlation plane metrics such as ONV, ACE, ASM and ACH. The performance of the OTMACH filter is improved by minimizing the energy function $E(\mathbf{h})$ of the correlation filter $\mathbf{h}$, given by

$$E(\mathbf{h}) = \alpha(ONV) + \beta(ACE) + \gamma(ASM) - \delta(ACH) \qquad (10.38)$$
$$= \alpha\mathbf{h}^+\bar{\mathbf{O}}\mathbf{h} + \beta\mathbf{h}^+\bar{\mathbf{D}}\mathbf{h} + \gamma\mathbf{h}^+\bar{\mathbf{S}}\mathbf{h} - \delta|\mathbf{m}^+\mathbf{h}|^2 \qquad (10.39)$$

These considerations lead to the expression for OTMACH filter as

$$\mathbf{h}_{\mathrm{OTMACH}} = \frac{\mathbf{m}}{\alpha\bar{\mathbf{O}} + \beta\bar{\mathbf{D}} + \gamma\bar{\mathbf{S}}} \qquad (10.40)$$

where $\alpha$, $\beta$ and $\gamma$ are the nonnegative optimal trade-off (OT) parameters.

## 10.4   Physical requirements in designing correlation filters

In physical terms, the correlation plane is treated as a new linearly transformed image generated by the filter in response to an input image. Therefore, not only the correlation peak but also the entire correlation plane needs to be tailored for better performance and hence ONV, ACE, ASM and

ACH have to be properly tuned with the help of parameters $\alpha, \beta, \gamma$ and $\delta$. In general, for face recognition applications, the values of non-negative constants $\alpha, \beta$ and $\gamma$ are chosen to tailor the filter's performance under noise and variations in illumination conditions and distortions in face images. The value of $\delta$ in minimizing the energy function in Equation 10.39 modifies the peak height at the correlation plane to ensure good correlation and therefore must dominate the other performance criteria. Minimization of ACE is required since the low value of ACE emphasizes the high frequency components of images. The control of trade-off parameters is possible in the OTMACH filter [161], which exhibits significantly better recognition performance than other filters. Easy detection of the correlation peak, better distortion tolerance and the ability to suppress the clutter noise are the three basic criteria that are fulfilled by using OTMACH. On the other hand, the OTSDF filter includes a trade-off parameter that takes a high value of $\alpha$ close to 1 and $\beta (= \sqrt{1 - \alpha^2})$ close to 0, so as to emphasize on the high value of ONV and low value of ACE. Similarly, the MVSDF filter is designed for minimum ONV but usually exhibits broad correlation peaks. Setting $\alpha = 0$, a MACE filter is designed. Though MACE produces a sharp peak, yet it is highly sensitive to noise and distortion and therefore its usefulness for robust face authentication in the presence of variations in the illumination condition is limited [162, 163]. Clutter rejection can be achieved by reducing the dependence on average training images by including ONV criteria. A tunable parameter $\beta$ is used to control the performance.

## MATLAB code for filter functions

```
%% All filter function

function H = Filter(A,alpha,beta,gamma,d1,d2)
d = d1*d2;
M = zeros(d,1);
S = zeros(d,1);
C = ones(d,1);
noI = size(A,3)

for ic = 1:noI
    f = A(:,:,ic);
    %f = double(f);
    f = imresize(f,[d1 d2]);
    F = fft2(f);
    M = M+F(:);
    S = S+F(:).*conj(F(:));
end
M = M./noI;
S = S./noI;
```

```
D = S;
S = S - M.*conj(M);

h = M./(alpha*C+beta*D+gamma*S);
H = reshape(h,d1,d2);
```

---

MATLAB code for PSR calculation

---

```
%% Function for PSR calculation
function [out] = PsrCalculation(Corr)
peak = max(max(Corr));
[h1,h2]= find(Corr==peak);
a = size(Corr,1);
b = size(Corr,2);
if h1>10 && h2>10 && h1<a-10 && h2<b-10
    Corr(ceil(h1-2):ceil(h1+2),ceil(h2-2):ceil(h2+2))=0;
    Mask = Corr(h1-10:h1+10,h2-10:h2+10);
    cnt = 1;
    for ic = 1:size(Mask,1)
        for ik= 1:size(Mask,2)
            if Mask(ic,ik) == 0

            else
                Annular(cnt,:)=Mask(ic,ik);
                cnt=cnt+1;
            end
        end
    end
    mn = mean(Annular);
    st = std(Annular);
    psr = (peak-mn)/st;
else
    psr=0;
end
out = psr;
```

---

## 10.5    Applications of correlation filters

Different constraint on the face images such as illumination variations, occlusion and expression variations yielded many types of correlation filters. In most of the cases, MACE, UMACE and their different phase extensions are used for verification purposes. Several noticeable works in face recognition using correlation filters can be found in [164, 165, 166, 163, 167, 168, 169, 170, 171, 172].

In [173], the MACE filter is synthesized with some training images and applied over the AMP facial expression database [174] where overall 0.1% equal error rate (EER) is achieved. A comparative performance of the MACE filter and individual eigenface subspace method (IESM) for face recognition in terms of margin of separations is presented in [175]. In [176] an efficient method of designing the MACE filter is proposed where the complexity of filters is reduced without sacrificing the system performance. Therefore even on limited resource platforms, the algorithm can perform face localization and recognition. An idea of incrementally updating the unconstrained filters for limited memory devices is also successfully proposed in [177] with incremental updating of a single training image one at a time. This updating method iteratively selects among the captured images during the enrollment stage. Boosting the performance of the MACE filter in illumination invariant face recognition using logarithmic transformation is proposed in [178]. It has been shown here that in using this transformation the MACE filter gives better discrimination between authentic and impostor PSRs. An approach to encrypting the MACE filter is reported in [179]. It has been shown here that an arbitrary random convolution kernel can be used. This helps to guard against the types of attacks where the attacker might try to intercept the decrypted filter during the verification stage.

In [180], a principal component analysis (PCA) is run on the phase spectrum of the training images in the Fourier domain allowing the phase information as a spanning linear subspace. The primary advantage of using a subspace to represent the target instead of a single filter is that it represents a larger set of target variations. This results in higher PSRs than the conventional MACE filter.

A successful combination of the support vector machine (SVM) with an advanced correlation filter to produce the maximum margin SVM correlation filter is proposed in [181]. It gives more control over the relationship of peaks to sidelobes in the training correlation planes. In addition, it inherently minimizes the sensitivity to additive white noise by minimizing filter energy subject to the existence of a margin.

Illumination invariant face recognition and impostor rejection using different minimum noise and correlation energy (MINACE) filter algorithms is proposed in [182]. Two different MINACE filter formulations, spectral envelope and additive spectrum, and two different correlation plane metrics,peak and peak-to-correlation plane energy ratio (PCER), were used to create face recognition systems that function with illumination variations. Good performance scores were presented for both face verification and identification on the PIE database.

A different approach to using the correlation filter is suggested in [183], where a quaternion array is developed from wavelet decomposition and used in synthesizing the correlation filter. By using the quaternion correlation filter to model the inter subband characteristics as well as the intra subband characteristics a decomposed representation is developed. The numerical experiments on the PIE data set show that this method achieves improvement when trained by a single near frontal lighting mug-shot image and tested on unknown, variable lighting face images. In the redundant class-dependence feature analysis (CFA) method for face recognition using correlation filters, the filters are designed, one for each subject, in the generic training set to get a bank of correlation filters. All these filters are used for feature extraction. The nearest neighbour rule is applied to decide on the class label for the test image.

Face class code (FCC)-based approach using the correlation filter and support vector machine (SVM) is proposed in [184]. This method is used as binary classifiers for face recognition when the number of the classes is large. FCC is combined with error control codes, and better recognition results under variable illumination conditions. The template matching method of correlation filter [185, 186] is used for facial feature extraction, where the cosine distance is measured from a similarity score. This method is successfully experimented over the FRGC2.0 dataset. In [187], it has been shown that kernel correlation feature analysis (KCFA) has good representation and discrimination ability for unseen datasets and produces better verification and identification rates on PIE, FERET and AR datasets.

In general, two dimensional (2D) correlation feature analysis (2DCFA) cannot be used for vectors and $N^{th}(N \geq 3)$ order tensors. This limitation is overcome by Yan et al. in [188], where a generalized method of analysis is proposed by using the image data as tensors. The improved recognition rate is obtained by the tensor-based method in comparison to traditional 2DCFA for standard face databases. An 1DCFA is proposed [189] in low-dimensional subspace (PCA) instead of 2DCFA, where peak height is minimized subject to linear constraint. Another kind of research in this area is called correntropy MACE (CMACE) filters [152]. In this case, the kernel function is limited to a Gaussian kernel. When combined with the fast Gaussian transform, the technique allows fast approximation of the full correlation output while retaining the increased representational power. In [152] it is shown that, although slow during operation, better face recognition rate is achieved with CMACE filters.

In [190] a comparative study of some recent advanced correlation filters is made to test recognition performance in different situations involving variations in facial expression, illumination conditions and head pose. It demonstrates that it is possible to obtain illumination invariance without using any training images for this purpose. The correlation filter classifier also has greater robustness and accuracy than traditional appearance-based methods (such as PCA). It has also reported that the phase extended unconstrained MACE filter is the best choice for facial matching.

Adaptive and robust correlation filters (ARCF) are proposed in [151] and describe their usefulness for reliable face authentication using recognition-by-parts strategies is described. ARCFs provide information that involves both appearance and location. The cluster and strength of the ARCF correlation peaks indicate the confidence of the face authentication made, if any. The adaptive aspect of ARCF comes from their derivation using both training and test data, similar to transduction, while the robust aspect benefits from the correlation peak optimization to decrease their sensitivity to noise and distortions.

An approach for face verification using local binary pattern (LBP) operators and optical correlation filters can be found in [191]. LBP is operated on training images to form local binary pattern-unconstrained minimum average correlation energy (LBP-UMACE) filters as an optical correlation filter to enhance recognition rates and reduce error rates simultaneously. Better performance of LBP-UMACE compared with UMACE filters is demonstrated.

In [192], the original idea based on the unconstrained optimal trade-off quaternion filter (UOTQF) is extended and two additional different correlation filters in quaternionic domain are evaluated: (1) a phase only quaternion filter (POQF) and (2) a separable trade-off quaternion filter (STOQF). Three different quaternion-based correlation filters are designed and conjugated with four face feature extraction methods. The advantage of synthesis correlation filters in quaternionic domain is only one face image of a person is needed for training. Combination of quaternionic representation with a quaternion-based correlation filter confirms good discriminating and illumination invariant properties and an improvement in face recognition accuracy is obtained.

An extended version of SVM and more generalized correlation filter approach is presented in [156] where the maximum margin correlation filter (MMCF) is proposed. It combines the generalization capability of SVM and localization capabilities of correlation filters. MMCF is successfully implemented in different object recognition and face classification problems.

## 10.6   Performance analysis

In this section some comparative performances of general purpose correlation filters are made for the face recognition task under various facial expressions and varying lighting conditions. Correlation filters like MACH, UMACE, OTMACH, quad phase UMACE (QPUMACE) and phase extended UMACE (PEUMACE) are synthesized and tested over several databases including AMP, Cropped YaleB, PIE and AR face datasets. The performance evaluation of different unconstrained correlation filters has been made by setting different optimal trade-off parameters in Equation 10.40. The UMACE filter is designed with the help of the equation, $h_{UMACE} = \bar{D}^{-1}m$, $(\alpha = 0, \beta = 1, \gamma = 0)$. Similarly MACH and OTMACH are designed using

$\mathbf{h}_{\text{MACH}} = \bar{\mathbf{S}}^{-1}\mathbf{m}, (\alpha = 0, \beta = 0, \gamma = 1)$ and $\mathbf{h}_{\text{OTMACH}} = (\alpha\bar{\mathbf{O}} + \beta\bar{\mathbf{D}} + \gamma\ bar\mathbf{S})^{-1}\mathbf{m}, (\alpha = 0.2, \beta = 0.5, \gamma = 0.3)$, respectively, where values of $\alpha, \beta, \gamma$ are chosen empirically. PEUMACE is obtained as the full phase extension of $\mathbf{H}_{\text{UMACE}}$ and is given by

$$\mathbf{H}_{\text{PEUMACE}} = e^{j\angle\mathbf{H}_{\text{UMACE}}} \tag{10.41}$$

In designing the QPUMACE filter each element in the filter array will take on $\pm 1$ for the real component. The imaginary component $\pm j$ is calculated in the following manner:

$$\mathbf{H}_{\text{QPUMACE}} = \begin{bmatrix} +1 & \Re\{\mathbf{H}_{\text{UMACE}}(u,v)\} \geq 0 \\ -1 & \Re\{\mathbf{H}_{\text{UMACE}}(u,v)\} < 0 \\ +j & \Im\{\mathbf{H}_{\text{UMACE}}(u,v)\} \geq 0 \\ -j & \Im\{\mathbf{H}_{\text{UMACE}}(u,v)\} < 0 \end{bmatrix} \tag{10.42}$$

PSR values are used to test verification accuracy of the above correlation filters for each database.

During performance analysis of correlation techniques for face recognition two types of tests are performed: (1) Identification test, where the class is labelled based on the filter that scores a relative maximum PSR and (2) Verification test, where an authentic test face images must achieve a score above a pre-set threshold. Face recognition performance of correlation filters is measured based on verification approach and it is maintained for every experiment on AMP, YaleB, PIE and AR face databases.

## 10.6.1   Performance evaluation using PSR values

All unconstrained filters are synthesized with the same number of training images (1,21,41) from Person-1 of the AMP database and tested over the whole database. Figure 10.6 shows the performance of different filters in terms of PSR values. The separation margin between authentic and impostor face images is calculated by subtracting the minimum PSR value of the authentic class and the maximum PSR value among the impostor classes. The largest distance of separation (DoS) is achieved for UMACE filter compared to others. UMACE filter is also tested where the training images (1,21,41) from Person-2 are used. From Figure 10.7 it is observed that the results degrade as reduced DoS is found for the UMACE. Hence the phase extension of the UMACE is considered since the phase contains more information than magnitude in an image. It is interesting to observe that while the full phase[2] instead of quad phase of UMACE is considered, DoS is increased indicating better performance for the face recognition task.

---

[2]The full phase extension of the test image is also taken during correlation.

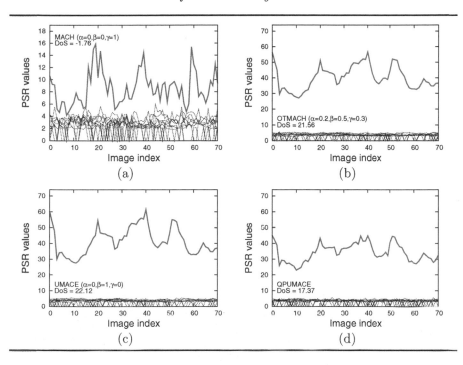

**FIGURE 10.6**: PSR performance of different unconstrained filter when tested over the AMP database and synthesized with Person-1.

## 10.6.2    Performance evaluation in terms of %RR and %FAR

Table 10.4 summarizes the %mean recognition rate with corresponding %false acceptance rate (FAR) while the correlation filter's performance is tested on the whole AMP database. In Table 10.4 the preset threshold is taken as 7. Table 10.5 shows the %mean recognition rate at zero FAR.

**TABLE 10.4**: The performance of different filters in face recognition on AMP facial expression database.

| Training images | MACH %rec,%far | UMACE %rec,%far | OTMACH %rec,%far | QPUMACE %rec,%far | PEUMACE %rec,%far |
|---|---|---|---|---|---|
| 1,21,41 | 90.124,0.9269 | 99.89,1.76 | 99.89,2.89 | 99.37,0.46 | 100,1.169 |
| 3,22,28 | 90.124,0.926 | 99.58,2.48 | 99.58,3.17 | 99.16,0.87 | 99.37,1.89 |
| 46,50,55 | 64.13,0 | 99.68,1.69 | 99.79,2.61 | 98.75,0.74 | 99.27,1.75 |

In case of performance evaluation of correlation filters on the Cropped

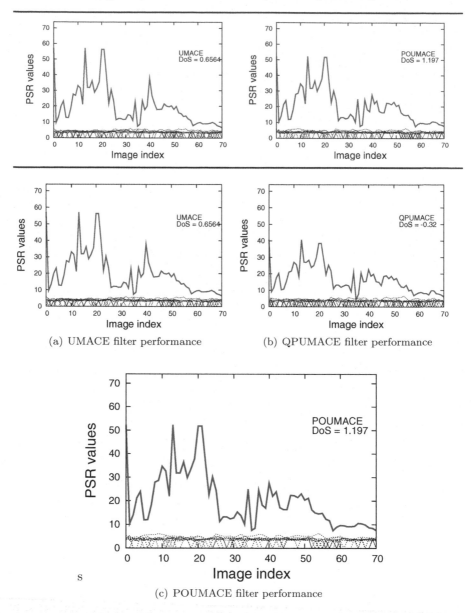

(a) UMACE filter performance    (b) QPUMACE filter performance

(c) POUMACE filter performance

**FIGURE 10.7**: PSR performance of UMACE and its full-phase extended variation. Filters are tested over the whole AMP database when synthesized with Person-2.

YaleB database, each filter is synthesized with each subset images and

**TABLE 10.5**: The %mean recognition rate on AMP database obtained by different filters when FAR = 0. %mean recognition rate increases when number of training images increases.

| Training images | MACH %rec | UMACE %rec | OTMACH %rec | QPUMACE %rec | PEUMACE %rec |
|---|---|---|---|---|---|
| 1,21,41 | 53.43 | 97.50 | 97.61 | 96.04 | 97.92 |
| 3,22,28 | 62.68 | 98.75 | 99.06 | 98.12 | 98.24 |
| 46,50,55 | 64.13 | 92.72 | 92.203 | 87.00 | 89.91 |
| 1,2,3,10,74 | 90.85 | 98.64 | 98.75 | 98.33 | 98.44 |

correlated over the whole database. Out of 38 persons a subset of 10 persons is taken for performance evaluation.

**TABLE 10.6**: The %mean recognition rate along with the %FAR obtained by different filters while the threshold is fixed at 10 (with no illumination compensation).

| Filters | UMACE %rec,%far | QPUMACE %rec,%far | PEUMACE %rec,%far | OTMACH %rec,%far |
|---|---|---|---|---|
| Subset-1 | 69.53,0.1042 | 71.1,0.0174 | 76.4,0.1215 | 89.22,16.99 |
| Subset-2 | 73.28,0.2431 | 69.37,0.086 | 73.90,0.257 | 89.06,21.2 |
| Subset-3 | 87.34,0.43 | 82.96,0.086 | 87.5,0.2431 | 94.68,10.42 |
| Subset-4 | 92.81,1.42 | 87.65,0.26 | 91.56,0.78 | 97.65,12.17 |
| Subset-5 | 92.18,2.06 | 73.9,0.69 | 82.03,1.54 | 98.59,18.92 |

Hence $10 \times 9 \times 64 (= 5760)$ number of impostor scores (PSR) and $10 \times 64 (= 640)$ authentic scores for YaleB are obtained. While testing with the PIE database $65 \times 64 \times 21 (= 87360)$ number of impostor scores and $65 \times 21 (= 1365)$ authentic scores for each filter are obtained. From the PSR distribution, %mean recognition rates are evaluated according to the verification method and the corresponding %FAR are recorded. Table 10.6 summarizes the %mean recognition rate of correlation filters while tested over the YaleB database. It is observed from Table 10.6 that the recognition results are greatly affected according to the choice of different lighting directions. Overall performance of the correlation filters shows that best recognition accuracy is obtained when subset-4 and subset-5 are chosen for training purpose. It is due to the fact these training sets have images with a wide variation of lighting or, in other words these images have illumination distributed evenly over the camera's visual field.

It is also observed that the OTMACH filter provides better %RR compared to others but from this result it cannot be concluded that OTMACH gives the best performance as %FAR is very high. Hence by considering both %RR and %FAR, it may be noted that the performance of PEUMACE in the case of subset-4 only is slightly better than the other filters.

---

MATLAB code for correlation filter for face recognition

---

```
%% Correlation filter for face recognition
%% UMACE, MACH, OTMACH etc..
close all
clear all
clc

ss1 =[1 7 8 9 37 38 36];
ss2 = [5 11 12 13 15 39 40 41 42 44 10 2];
ss3 =[3 6 14 16 17 19 20 45 48 49 43 46];
ss4 = [18 21 22 23 24 25 26 50 51 52 53 54];
ss5 =[4 35 29 30 31 32 33 34 28 27 64 63 62 61 56 57 58 59 60];

for ic = 1:64
    alltest(ic)=ic;
end

trainSet = ss5;
testSet = alltest;

d1 = 100; d2 = 100;
PsrUmace = zeros(size(testSet,2),10);
PsrMach = zeros(size(testSet,2),10);
PsrOtmach = zeros(size(testSet,2),10);
for class = 1:10

    for ic = 1:size(trainSet,2)
        A(:,:,ic) = imread(strcat("yaleB",num2str(class),...
        "_",num2str(trainSet(:,ic)),".pgm"));
    end

    Umace = Filter(A,0,1,0,d1,d2);
    Mach = Filter(A,0,0,1,d1,d2);
    Otmach = Filter(A,0.9,0.9,0.8,d1,d2);


    for ic = 1:size(testSet,2)
        t = imread(strcat("yaleB",num2str(class),...
        "_",num2str(testSet(:,ic)),".pgm"));
        %t = double(t);
        t = imresize(t,[d1 d2]);
        T = fft2(t);
```

```
            corr = real(fftshift(ifft2(T.*conj(Umace))));
            psr = PsrCalculation(corr);
            PsrUmace(ic,class) = psr;
            clear corr psr
            corr = real(fftshift(ifft2(T.*conj(Mach))));
            psr = PsrCalculation(corr);
            PsrMach(ic,class) = psr;
            clear psr corr
            corr = real(fftshift(ifft2(T.*conj(Otmach))));
            psr = PsrCalculation(corr);
            PsrOtmach(ic,class) = psr;

    end
end
AvgPsrUmace = mean(PsrUmace,2);
AvgPsrMach = mean(PsrMach,2);
AvgPsrOtmach = mean(PsrOtmach,2);

% Class specific PCA
for trainPerson=1:10
    T=[];
    No_of_Training_Images=size(trainSet,2); % for each class
    C=1;
    for h=1:No_of_Training_Images
            hh=int2str(trainPerson);
            kk=int2str(trainSet(:,h));
            b=strcat("yaleB",hh,"_",kk);
            img=imread(strcat(b,".pgm"));
            f = double(img);
            f = imresize(f,[100 100]);
            F = fft2(f);
            Df = F(:).*conj(F(:));
            hf = F(:)./Df;
            Pf = exp(1j.*angle(hf));
            [m1,n1]=size(f);

            T=[T Pf];
            C=C+1;
    end
    % Number of classes (or persons)
    Class_number = ( size(T,2) )/(C-1);
    % Number of images in each class
    Class_population = C-1;
    % Total number of training images
    P = Class_population * Class_number;

    % figure(1)
    m_total=mean(T,2);
    Mimg=reshape(m_total,m1,n1);
```

```
%imshow(Mimg,[]);title("MEAN IMAGE");

Difference=[];
for ic=1:size(T,2)
    diff=T(:,ic)-m_total;
    Difference=[Difference diff];
end

Covar=Difference"*Difference;
[U,E,V]=svd(Covar);
val=diag(E);

figure(3)
stem(val);title("EIGEN VALUE");
drawnow;
Eigen_Vector=Difference*U;
Eigen_Vector=U;
figure(4)
for ic=1:size(U,2)
    Eigen_Face=Eigen_Vector(:,ic);
    Eigen_Face_Image=reshape(Eigen_Face,m1,n1);
    subplot(ceil(sqrt(size(U,2))),ceil(sqrt(size(U,2))),ic);
    imshow(Eigen_Face_Image,[]);
    drawnow;
 end

PC=Eigen_Vector;
for ic=1:size(PC,2)
    PC(:,ic)=PC(:,ic)./norm(PC(:,ic));
end

%%% Weight calculation of Training Images

ProjectedImages_PCA = [];
for ic = 1 : P
    temp = transpose(PC)*Difference(:,ic);
    ProjectedImages_PCA = [ProjectedImages_PCA temp];
end

% Reconstruction by PCA
clear f

testPerson = trainPerson;
for ic = 1:size(testSet,2) %person index
    hh=int2str(testPerson);
    k = testSet(:,ic);
    kk=int2str(k);
    b=strcat("yaleB",hh,"_",kk);
```

```
f=imread(strcat(b,".pgm"));

f=double(f);
f= imresize(f,[100 100]);
F = fft2(f);
Df = F(:).*conj(F(:));
hf = F(:)./Df;
Pf = exp(1j.*angle(hf));
        figure(1)
        imagesc(L),colormap(gray);
        figure(1)
        imagesc(L),colormap(gray); title("Test Image");
diff=Pf-m_total;
projected=transpose(PC)*diff;
reconstructed=m_total+PC*projected;
Recn=reshape(reconstructed,m1,n1);
Pf = reshape(Pf,[100 100]);

corr = real(fftshift(ifft2(Pf.*conj(Recn))));
%figure, surf(corr);view([-34 10])
psr = PsrCalculation(corr);
PSR(ic,testPerson) = psr;

    end
end
AvgPsrCsPca = mean(PSR,2);
AvgPSR = [AvgPsrUmace AvgPsrMach AvgPsrOtmach AvgPsrCsPca];
plot(AvgPSR),axis([0 64 0 100])
```

MATLAB code for Coreface

```
%% Coreface matlab program

clear all
close all
clc
%% NO NORMALIZATION HAVE BEEN DONE

% dir = "F:\croppedyale";
% cd(dir);
PSR = zeros(64,10);
o =imread("NLP1_1.jpg");
%imagesc(o);colormap(gray)

ss1 =[1 7 8 9 37 38 36];
ss2 = [5 11 12 13 15 39 40 41 42 44 10 2];
ss3 =[3 6 14 16 17 19 20 45 48 49 43 46];
ss4 = [18 21 22 23 24 25 26 50 51 52 53 54];
ss5 =[4 35 29 30 31 32 33 34 28 27 64 63 62 61 56 57 58 59 60];

ss6 = [ 1 5 3 18 4];
ss7 =[1 4 40 54 25];
for Knownclass=1:1
    Knownclass
    T=[];
    No_of_Training_Images=size(ss1,2); % for each class
    C=1;
        for h=1:No_of_Training_Images
            hh=int2str(Knownclass);
            kk=int2str(ss1(:,h));
            b=strcat("yaleB",hh,"_",kk);
            img=imread(strcat(b,".pgm"));
            f = double(img);
            f = imresize(f,[100 100]);
            F = exp(1j.*angle(fft2(f)));
            [m1,n1]=size(f);
            figure(1);
            subplot(ceil(sqrt(12)),ceil(sqrt(12)),C);
            imshow(img);
            if h==3
                title("TRAINING IMAGES");
            end
            T=[T F(:)];
            C=C+1;
        end
```

```
Class_number = ( size(T,2) )/(C-1);
Class_population = C-1;
P = Class_population * Class_number;
m_total=mean(T,2);
Mimg=reshape(m_total,m1,n1);
%imshow(Mimg,[]);title("MEAN IMAGE");

Difference=[];
for ic=1:size(T,2)
    diff=T(:,ic)-m_total;
    Difference=[Difference diff];
end

Covar=transpose(Difference)*Difference;
[U,E,V]=svd(Covar);
val=diag(E);

figure(3)
stem(val);title("EIGEN VALUE");
drawnow;
Eigen_Vector=Difference*U;
 Eigen_Vector=U;
 figure(4)
 for ic=1:size(U,2)
    Eigen_Face=Eigen_Vector(:,ic);
    Eigen_Face_Image=reshape(Eigen_Face,m1,n1);
    subplot(ceil(sqrt(size(U,2))),ceil(sqrt(size(U,2))),ic);
    imshow(Eigen_Face_Image,[]);

    drawnow;
 end

PC=Eigen_Vector;
for ic=1:size(PC,2)
    PC(:,ic)=PC(:,ic)./norm(PC(:,ic));
end
%% Weight calculation of Training Images

ProjectedImages_PCA = [];
for ic = 1 : P
    temp = transpose(PC)*Difference(:,ic);
    ProjectedImages_PCA = [ProjectedImages_PCA temp];
end

%Reconstruction by PCA

person = 1;
PSR = [];
  testSet = ss2;
```

```
for  ic = 1:size(testSet,2) %person index
    hh=int2str(person);
    k = testSet(:,ic);
    kk=int2str(k);
    b=strcat("yaleB",hh,"_",kk);
    f=imread(strcat(b,".pgm"));
    f=double(f);
    f= imresize(f,[100 100]);
    F = exp(1j.*angle(fft2(f)));
    %          figure(1)
    %          imagesc(L),colormap(gray);
    %          figure(1)
    %          imagesc(L),colormap(gray); title("Test Image");
    diff=F(:)-m_total;
    projected=transpose(PC)*diff;
    reconstructed=m_total+PC*projected;
    Recn=reshape(reconstructed,m1,n1);
    corr = real(fftshift(ifft2(Recn.*conj(F))));
    %figure,surf(corr);view([-34 10])
    psr = PsrCalculation(corr);
    PSR = [PSR;psr];

    end
  end
PSR
```

### 10.6.3    Performance    evaluation    by    receiver    operating characteristics (ROC) curves

Another way of observing the performance is by plotting receiver operating characteristics (ROC) curves. The performance of correlation filters can be characterized in terms of the probabilities of correct detection ($P_D$) and probability of false alarm ($P_{FA}$). In general low detection thresholds improve the probability of correct recognition, while large thresholds decrease false alarm probabilities by rejecting erroneous peaks (or specifically PSRs). The relationship of $P_D$ and $P_{FA}$ with threshold PSR can be represented by ROCs. ROCs are calculated with increasing PSRs as thresholds. When comparing ROC curves of different tests, curves for better performance lie closer to the top left corner and the worst case performance is indicated by a diagonal line. The diagonal line represents $P_D = P_{FA}$. The curves nearer to the diagonal line represent the worst detection performance. Figure 10.8 shows that in general the average filter performance is best when subset-4 is used as a training set. This can be explained, as subset-4 includes the training images having wide illumination variation compared to others. Hence any face image that lies in the convex hull of these training images should be perfectly recognized.

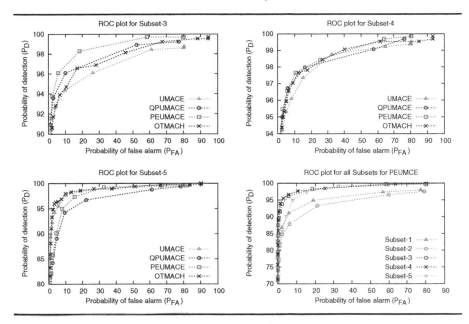

**FIGURE 10.8**: ROC plots of different correlation filters for different subsets. The performance of the PEUMACE filter is also observed for different subset training.

## 10.7 Video correlation filter

Automatic detection of targets is the first step in most automatic vision systems. A face in an input video scene can be considered as a target. Hence in automatic face detection and recognition an input scene is similar to automatic target detection/recognition (ATD/R) problem. Correlation filters are well suited to such applications due to their attractive properties such as shift-invariance, distortion tolerance and closed-form solutions. They have been used successfully for target detection and recognition in scenes with unknown numbers of targets and heavy clutter [193, 194].

Conventional correlation filtering may be divided into two stages: the design or synthesis stage and the test stage. The design stage (offline process) of a correlation filter is often computationally expensive; however, each filter need only be synthesized once, and using (online process) the filter thereafter can be done efficiently in the frequency domain. In most correlation filter-based ATD/R systems, each input frame is first passed through one or more filters, and peak locations are then identified and the rest of the information in the output is discarded. Hence this type of ATD/R system is a frame-based approach demanding high computational cost.

Another drawback of frame-based approach in face detection is that the temporal information, the most important part of a video scene, is ignored. In this study a face detection method by fully using the temporal information provided by video is described. That is instead of detecting each frame, the temporal approach exploits temporal relationships between the frames to detect multiple human faces in a video sequence. In this context a modified version [195] of the UMACE filter is generalized to a video or a 3D spatiotemporal volume, termed an unconstrained video filter (UVF). After correlating this UVF with the target video a probable location of face is detected according to the position of high correlation peak in a three dimensional plane. This location is the region of interest (ROI) in the target scene. The ROI is extracted and fed to the DCCF [196] for classification.A Detailed process of face detection and recognition system is shown in Figure 10.9. The effectiveness of this automatic system is simulated on the VidTIMIT audio-video database [197].

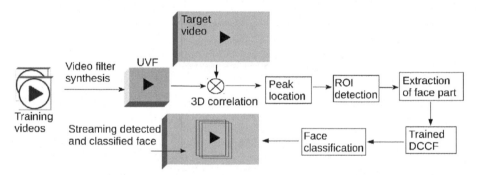

**FIGURE 10.9**: Detailed process of face detection and recognition in a video using synthesized correlation filters.

## 10.8    Formulation of unconstrained video filter

A 3D spatiotemporal volume corresponding to a maximum average correlation height (MACH) filter is proposed in [153] for action recognition in videos. Better intra-class tolerance can be achieved by the MACH filter as it maximizes the average correlation height (ACH) while minimizing average similarity measure (ASM). Hence the MACH filter is useful for recognition purposes where intra-class variation is dominant. But for a generalized face detection system exact location of face is needed instead of recognizing a specific class of action. Another important thing is that while using the correlation filters in detection purposes the peak location in the correlation plane is searched. For exact location sharp and distinct peak is needed and that can be achieved by using the MACE type filter where minimization of correlation energy is achieved by suppressing side-lobes around the peak. In this study a modified unconstrained optimal trade off SDF (MUOTSDF) filter is elaborated. At first the 2D version of this filter is given and then it is generalized to a 3D spatio-temporal volume or video filter.

## 10.8.1 Mathematical formulation of MUOTSDF

The general UMACE filter solution is given in Equation 10.43. For robust face recognition, FAR and FRR are important issues. When selecting a correlation filter for face recognition purpose these issues need to be addressed. In most cases of frequency domain correlation techniques it is observed that the MACE or the UMACE type filters perform efficiently, as these filters amplify the high frequency components. This helps the filters to recognize face images under different lighting conditions as the edges of faces are greatly enhanced. Moreover, for face recognition, it is desired to get a sharp correlation peak and simultaneously the filter must be sensitive to the distortion. If the face data base contains only frontal face images under different lighting conditions, where no variations in pose or expression is allowed, the UMACE solution of OTMACH parameter can be considered by neglecting $\gamma$. Hence the general OTMACH solution can be reduced to

$$\mathbf{h}_{\text{UMACE}} = \mathbf{D}^{-a}(\bar{\mathbf{D}}^{-b}\mathbf{m}) \qquad (10.43)$$

where $(a + b) = 1$. The computation of $\mathbf{h}_{\text{UMACE}}$ is trivial as $\bar{\mathbf{D}}$ is a diagonal matrix constructed by averaging the power spectra of training face images. The pre-whitening spectrum stage $a$ will emphasize the high frequency components while suppressing the low frequency components which are responsible for illumination variations in face images. Illumination variations in face images are mostly reflected in the lower frequency spectrum of the FFT of face images. Hence the value of the weight $a$ is chosen in such a way so as to give good tolerance to illumination variations. The phase matching weight $b$ basically denotes a feature of correlation matching. The correlation filters like UMACE have an in-built property of recognizing the face images under poor illumination conditions. But to produce sharp discernible correlation peaks the UMACE filter emphasizes high frequencies and can result in poor intraclass recognition of images not included in the training set. Hence to perform classification using a simple UMACE filter solution sometimes increases both FAR and FRR. Therefore, it is necessary to modify the UMACE formulation to improve upon the performance under varying lighting conditions. A metric can be introduced in order to force the correlation outputs from all images in the training set to match the average of the correlation outputs from some exemplars. In the case of MACH and UMACE filters, conventionally, $\mathbf{x}_i$ is used as an exemplar. However, instead of using $\mathbf{x}_i$, $(\mathbf{x}_i - \beta\mathbf{m})$ is introduced to modify the UMACE filter solution so that the relative influence of average image is incorporated in the filter solution. Here $\beta$ is the controlling parameter depending on what relative influence of the mean image is exploited. The exemplar $(\mathbf{x}_i - \beta\mathbf{m})$ is now the $i$th training image with part of the mean subtracted. Hence it is desirable for all images in the training set to follow these exemplars' behaviours. This can be done by forcing every image in the training set to have a similar correlation output plane to an ideal correlation output shape $\mathbf{f}$. To find the $\mathbf{f}$ that best matches all these exemplars' correlation output planes its deviation from their correlation plane is minimized. This deviation can be quantified by the average squared error (ASE) as

$$\text{ASE} = \frac{1}{N}\sum_{i=1}^{N}|\mathbf{g}_i - \mathbf{f}|^2$$

$$= \frac{1}{N}\sum_{i=1}^{N}(\mathbf{g}_i - \mathbf{f})^+(\mathbf{g}_i - \mathbf{f}) \tag{10.44}$$

where

$$\mathbf{g}_i = (\bar{\mathbf{X}}_i - \beta\bar{\mathbf{M}})^*\mathbf{h} \tag{10.45}$$

where $\bar{\mathbf{X}}_i = diag\{\mathbf{x}_i\}$ and $\bar{\mathbf{M}} = diag\{\mathbf{m}\}$. Equation 12.13 represents the correlation plane in vector form in response to the $i$th training image. To find the optimum shape vector $\mathbf{f}^{opt}$ the gradient of ASE in Equation 12.12 is set to zero and $\mathbf{f}^{opt}$ is obtained as

$$\nabla_f(\text{ASE}) = \frac{2}{N}\sum_{i=1}^{N}(\mathbf{g}_i - \mathbf{f}) = 0 \tag{10.46}$$

or

$$\mathbf{f}^{opt} = \frac{1}{N}\sum_{i=1}^{N}\mathbf{g}_i \tag{10.47}$$

Hence the optimal shape vector can be formulated as

$$\mathbf{f}^{opt} = \frac{1}{N}\sum_{i=1}^{N}(\bar{\mathbf{X}}_i - \beta\bar{\mathbf{M}})^*\mathbf{h}$$

$$= \left\{\frac{1}{N}\sum_{i=1}^{N}\bar{\mathbf{X}}_i - \beta\bar{\mathbf{M}}\right\}^*\mathbf{h}$$

$$= \{(1 - \beta)\bar{\mathbf{M}}\}^*\mathbf{h} \tag{10.48}$$

Now the average similarity measure can be modified as the measure of dissimilarity of the training images to $(1-\beta)\bar{\mathbf{M}}^*\mathbf{h}$ and can be mathematically expressed as

$$\text{ASM}_{new} = \frac{1}{N}\sum_{i=1}^{N}|\bar{\mathbf{X}}_i^*\mathbf{h} - (1-\beta)\bar{\mathbf{M}}^*\mathbf{h}|^2$$

$$= \mathbf{h}^+\bar{\mathbf{S}}_{new}\mathbf{h} \qquad (10.49)$$

where

$$\bar{\mathbf{S}}_{new} = \frac{1}{N}\sum_{i=1}^{N}(\bar{\mathbf{X}}_i - (1-\beta)\bar{\mathbf{M}})(\bar{\mathbf{X}}_i - (1-\beta)\bar{\mathbf{M}})^* \qquad (10.50)$$

It is considered here that images are corrupted with additive white Gaussian noise, which is the most common noise model. The noise is characterized by its variance, which is an important parameter for the majority of image denoising algorithms, because it controls the strength of the filtering. Hence, in addition to ASM, MUOTSDF design includes the ONV term. As the ONV estimates the variance of the noise of a correlation plane, here it is also minimized along with the ASM term. Minimizing both ONV and ASM (given in Equation 12.18 terms and maximizing the ACH, the optimal filter MUOTSDF can be expressed by redrawing Equation 10.43 by replacing $\bar{\mathbf{D}}$ by $\alpha\bar{\mathbf{O}} + \beta\bar{\mathbf{S}}_{new}$ as

$$\mathbf{h}_{\text{MUOT3DF}} = (\alpha\bar{\mathbf{O}} + \beta\bar{\mathbf{S}}_{new})^{-a}(\alpha\bar{\mathbf{O}} + \beta\bar{\mathbf{S}}_{new})^{-b}\mathbf{m} \qquad (10.51)$$

where $\alpha$ and $\beta$ are scalar parameters and $a + b = 1$.

## 10.8.2 Unconstrained video filter

A simple and straightforward way to locate the face in a single video frame can be obtained by simple correlation of $\mathbf{h}_{\text{MUOTSDF}}$, given in Equation 10.51, with the successive 2D video templates. But in order to fully encompass the information of both space and time contained in a video sequence, $\mathbf{h}_{\text{MUOTSDF}}$ is generalized to UVF. UVF is synthesized by the information obtained from spatiotemporal volumes of consecutive face video sequences. A series of spatio-temporal volumes i.e. some video files, are taken from the face video sequences and concatenated with the frames of a single complete cycle to synthesize UVF. From a set of spatio-temporal volumes the temporal derivatives of each pixel of each video sequence are calculated by the Sobel operator [198].

It is a differential operator computing an approximation of the gradient of the image intensity and it is very fast to apply since it is based on a small window ($3\times3$ kernel) to convolve with the whole image. Equation 10.52 shows the temporal derivative operation of one video sequence with the Sobel kernel. Two matrices are referenced here, one for the gradiant over the x-axis and one for the gradiant over the y-axis. Hence for a given image, convolution is made two times in order to obtain $E_x$ and $E_y$ that are the images containing the approximation of spatial derivatives.

$$E_x = F(x,y) * \frac{1}{8} \begin{pmatrix} +1 & +2 & +1 \\ 0 & 0 & 0 \\ -1 & -2 & -1 \end{pmatrix} \quad , \quad E_y = F(x,y) * \frac{1}{8} \begin{pmatrix} +1 & 0 & -1 \\ +2 & 0 & -2 \\ +1 & 0 & -1 \end{pmatrix}$$

(10.52)

where $F(x,y)$ is one of the frames in the video scenes.

The coefficient $\frac{1}{8}$ has the purpose to smooth the derivative in such a way that the peaks of the derivative function are lowered. Final edge image $E(x,y)$ can be obtained as

$$|E(x,y)| = |E_x| + |E_y|$$

(10.53)

These edge images ($E(x,y)$) are then stored in a 3D matrix to construct the spatio-temporal volumes and the set of spatio-temporal volumes are then processed in the frequency domain by 3D FFT for further synthesizing the UVF. The 3D FFT operation of the spatio-temporal volume is given by

$$\mathbf{E}_{3D}(u,v,w) = \sum_{t=0}^{T-1} \sum_{y=0}^{C-1} \sum_{x=0}^{R-1} \hat{E}(x,y,t) e^{-j2\pi[\frac{ux}{R} + \frac{vy}{C} + \frac{wt}{T}]}$$

(10.54)

where $\mathbf{E}_{3D}(u,v,w)$ is the resulting volume in the frequency domain obtained from the volume $\hat{E}(x,y,t)$ corresponding to the temporal derivative of the input sequence. $C$ is the number of columns, $R$ the number of rows and $T$ the number of frames in one video training set. Having obtained the volume in the frequency domain, the 3D matrix $\mathbf{E}_{3D}(u,v,w)$ is lexicographic ordered and the resulting column vector $\mathbf{e}_i$ of dimension $T \times C \times R$ (where $i = 1,2,3,\cdots T_s$). $T_s$ is the total number of frames used and the whole training set is obtained. From a set of $\mathbf{e}_i, i = 1,2,...,N$, $\mathbf{h}_{\text{MUOTSDF}}$ is synthesized.

Having obtained the 1D filter $\mathbf{h}_{\text{MUOTSDF}}$ it is now reshaped in the reverse order by arranging the vector elements into a volume containing $R$ rows, $C$ columns and $T$ frames with proper care. Thus a 3D MUMACE or unconstrained video filter (UVF) $\mathbf{H}_{3D}$ is generated and the corresponding spatial domain filter $H_{3D}$ can be obtained by 3D inverse Fourier transform according to the following equation,

$$H_{3D}(x,y,t) = \sum_{u=0}^{R-1} \sum_{v=0}^{C-1} \sum_{w=0}^{T-1} \mathbf{H}_{3D}(u,v,w) e^{j2\pi[\frac{ux}{R} + \frac{vy}{C} + \frac{wt}{T}]}$$

(10.55)

where $x = 0,1,2,\cdots,R-1,\quad y = 0,1,2,\cdots,C-1,\quad t = 0,1,2,\cdots,T-1$. Figure 10.10 shows the volumetric representation of the UVF filter in gray scale. This UVF $\mathbf{H}_{3D}$ is correlated with video clips and the face part is extracted. This face image is further verified with DCCF.

MATLAB code for uncostrained video filter

```
%% Unconstrained Video Filter

clear all
close all
clc
warning off
numVolumes = 20;
volumes = cell(1, numVolumes);

figure(1); cn=1;
for v = 1: 20%numVolumes
    inFile = sprintf("FTrain%d.avi", v);
    ifp = aviinfo(inFile);
    volume = zeros(ifp.Height, ifp.Width, ifp.NumFrames, "uint8");
    for f = 1 : ifp.NumFrames
        frame = aviread(inFile, f);
        rgbImg = frame.cdata;
        grayImg = rgb2gray(rgbImg);
        edgeImg = sobel(grayImg);
        volume(:,:,f) = edgeImg;
        imshow(volume(:,:,f));
        pause(0.0000002)
    end
    volumes{cn} = volume;
    cn=cn+1;
end

%% Make 3D Filter

[imgRows imgCols timeSamples] = size(volumes{1});
d = imgRows * imgCols * timeSamples;
N = length(volumes);
x = zeros(d, N);
for i = 1 : N
    fft_volume = fft3(double(volumes{i}),[imgRows  imgCols  timeSamples]);
    x(:,i) = fft_volume(:);
end
clear volumes;
mx = mean(x, 2);
c = ones(d,1);  % 2 * ones(d,1);
dx = mean(conj(x) .* x, 2);
temp = x - repmat(mx, 1, N);
sx = mean(conj(temp) .* temp, 2);

alpha = 0.9%0.1%0.01;  % 0.05; 1e-3; %0.05; % 0.01;
```

```
beta =   0.0000000000000009 % 1e-15; % 1e-12;       % 0.3
gamma = 0.00000000000000006; % 1e-12;   0.1;
h_den = (alpha * c) + (beta * dx) + (gamma * sx);
h = mx ./ h_den;
h = reshape(h, [imgRows, imgCols, timeSamples]);
h = real(ifft3(h));
h = uint8(scale(h, min3(h), max3(h), 0, 255));
UVF = h;
save UVF.mat UVF
%% Save 3D MACH as a short movie clip

outFile = "UVF.avi";
mov = avifile(outFile, "COMPRESSION", "None", "FPS", ifp.FramesPerSecond,
"QUALITY", 100);
% "Indeo5" is better and offer more compression than "Cinepak"
figure(2);
for f = 1 : ifp.NumFrames
    rgbMACH = cat(3, UVF(:,:,f), UVF(:,:,f), UVF(:,:,f));
    m = im2frame(rgbMACH);
    mov = addframe(mov, m);
    imshow(rgbMACH);
    pause(0.08);
end
mov = close(mov);
clear c
```

## 10.9    Distance classifier correlation filter

A detailed mathematical description of distance classifier correlation filter (DCCF) can be found in [196]. The MACE type filter is not well known for its ability to handle distortions and is also very much sensitive to intra-class variations. This is due to the fact the MACE filter design does not include any class compactness approach. Unlike the MACE filter the DCCF design maximizes the distance of all classes (= L) from the central mean by formulating the measure $\mathbf{h}^+\bar{\mathbf{M}}\mathbf{h}$, where

$$\bar{\mathbf{M}} = \frac{1}{L}\sum_{k=1}^{L}(\mathbf{m} - \mathbf{m}_k)(\mathbf{m} - \mathbf{m}_k)^+ \tag{10.56}$$

where $\mathbf{m}_k$ is the $k$th class mean and $\mathbf{m}$ represents global mean of entire training set. The individual correlation peaks of the different classes are also well separated, provided the criterion for compactness given by $\mathbf{h}^+\bar{\mathbf{S}}\mathbf{h}$, where

$$\bar{\mathbf{S}} = \frac{1}{L}\sum_{k=1}^{L}\frac{1}{N}\sum_{i=1}^{N}\{\bar{\mathbf{X}}_{ik} - \bar{\mathbf{M}}_k\}\{\bar{\mathbf{X}}_{ik} - \bar{\mathbf{M}}_k\}^* \qquad (10.57)$$

is simultaneously minimized. As the design process includes the average similarity measure (ASM) term $\mathbf{h}^+\bar{\mathbf{S}}\mathbf{h}$, designed $k$th class DCCF becomes robust in recognizing $k$th class faces while rejecting other classes. Due to the distortion tolerance ability DCCF is included in the face classification stage. The optimum solution of DCCF is the dominant eigenvector of $\bar{\mathbf{S}}^{-1}\bar{\mathbf{M}}$. For testing purposes the distance to be computed is

$$d_k = |\mathbf{H}^*\mathbf{z} - \mathbf{H}^*\mathbf{m}_k|^2 = p + b_k - 2\mathbf{z}^+\mathbf{h}_k \qquad (10.58)$$

where $\mathbf{z}$ is the input image, $p = |\mathbf{H}^*\mathbf{z}|^2$ is the transformed input image energy, $b_k = |\mathbf{H}^*\mathbf{m}_k|^2$ is the energy of the transformed $k$th class mean and $\mathbf{h}_k = \mathbf{H}\mathbf{H}^*\mathbf{m}_k$. The target is labeled to the class for which $d_k$ in Equation 10.58 is found to be minimum.

---

## 10.10    Application of UVF for face detection

### 10.10.1    Training approach

The video of each person is stored as JPEG images with a resolution of $512 \times 384$ pixels. Before developing the video files, all images are re-sized to $128 \times 128$. For each person the image sequences corresponding to the first two sentences are used to develop the training video files. Initially face parts are cropped from original video sequences. From these cropped faces, video files are generated with 20 frames per second. Each training video is of 2-second length. Figure 10.10 shows the step by step training approach of making the UVF. This UVF is designed for detecting the face in the target scene. After detecting the face, classification is made by DCCF. Hence training of DCCF is needed. The cropped face images are used to train DCCF. Each cropped face is resized to $64 \times 64$ for further DCCF training. For a given database of $L$ number of classes, $L$ number of DCCFs will be generated as shown in Figure 10.11.

### 10.10.2    Testing approach

For each person out of ten sentences, images with the first two sentences are used for training and, with the images corresponding to the rest of the sentences eight test video files are developed. Hence $43 \times 8$ test video files are

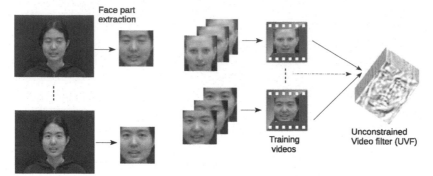

**FIGURE 10.10**: Detailed training process and volumetric representation of UVF

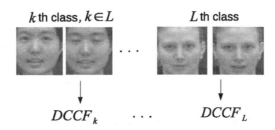

**FIGURE 10.11**: Training of DCCFs

generated. Test videos are developed with full $128 \times 128$ image sequences and have different time lengths.

---

MATLAB code for DCCF for multiclass pattern recognition

---

```
%% DCCF for multiclass pattern recognition

close all
clear all
clc

NoC = 3;
NoI = 14;
M = zeros(4096,NoC);
D = M;
S = M;
ic=1;
for ih = 1:NoC
    for ik = ih*14-14+1:ih*14-14+NoI
        f = imread(strcat(num2str(ik),".jpg"));
```

```
        %f = im2double(f);
        imshow(f);pause(0.2)
        F = fft2(f);
        M(:,ic) = M(:,ic) + F(:);
        D(:,ic) = D(:,ic) + F(:).*conj(F(:));
    end
    M(:,ic) = M(:,ic)/NoI;
    S(:,ic) = D(:,ic)./NoI;
    S(:,ic) = S(:,ic)-M(:,ic).*conj(M(:,ic));
    ic = ic+1;
end

Stotal = zeros(4096,1);
Mtotal = Stotal;
for ic = 1:NoC

    Stotal = Stotal+S(:,ic);
    Mtotal = Mtotal + M(:,ic);
end

Stotal = Stotal/NoC;
Mtotal = Mtotal/NoC;

% Formulating E
E =[];
for ic = 1:NoC
    e = Mtotal - M(:,ic);
    E = [E e];
end

% Formulating V
V = (conj(E))'*E;

% Eigens of V

[P,val,Q]=svd(V);

% Calculating Phi
Phi = E*P*(val)^(-0.5);

% calculating a
term =[];
for ic = 1: size(Phi,2)
    t = Phi(:,ic)./Stotal;
    term = [term t];
end

temp = val*transpose(conj(Phi))*term;
[a,lambda,a1] = svd(temp);
```

```
amax = a(:,1);

% Calculating h
h = (Phi*amax)./Stotal;

for ic = 1:NoC

    bf = (M(:,ic).*conj(h));
    b(:,ic) = transpose(bf)*bf/4096;
end
for ic = 1:NoC
    hf(:,ic) = h.*conj(h).*M(:,ic);
    H(:,:,ic) = reshape(hf(:,ic),64,64);
end
%save DCCF.mat H h b

Dist =[];
r=1;
c=1;
for ih = 1:NoC
    for ik = ih*14-14+1:ih*14-14+14
        test = imread(strcat(num2str(ik),".jpg"));
        imshow(test,[]);pause(0.02);
        ftest = fft2(test);
        z = ftest(:);

        p = (z.*conj(h));
        p = transpose(p)*p/4096;

        for ic = 1:NoC
            g = real(fftshift(ifft2(ftest.*conj(H(:,:,ic)))));
            d = p + b(:,ic) - 2* max(g(:));

            Dist =[Dist d];
        end
        indx = find(Dist==min(Dist));
        Index(r,c) = indx;
        r = r+1;
        Dist=[];
    end
    c=c+1;
    r=1;
end
Index
```

### 10.10.3 Face detection in video using UVF

From test videos the frames of dimension $128 \times 128$ are extracted and converted to gray-scale images. These gray-scale images are convoluted with the Sobel kernel for temporal derivative operations. The changes of direction in facial parts like lips, mouth and eyes during reciting a sentence become prominent in edge images after this convolution operation. Edge images are then stored in a 3D matrix according to the optical flow of video stream. The volumetric representation of edge image 3D matrix is shown in Figure 10.12. This volume is then Fourier transformed by the 3D FFT algorithm. The Fourier transformed volume is then correlated with the designed 3D UVF. Instead of producing a correlation plane this 3D correlation approach results in a correlation volume as shown in Figure 10.12. Instead of performing multi-correlation framewise (for a video length of 128 frames, 128 correlations are needed), which will take much time, 3D correlation is performed. The usefulness of 3D correlation is that it is a time saving approach.

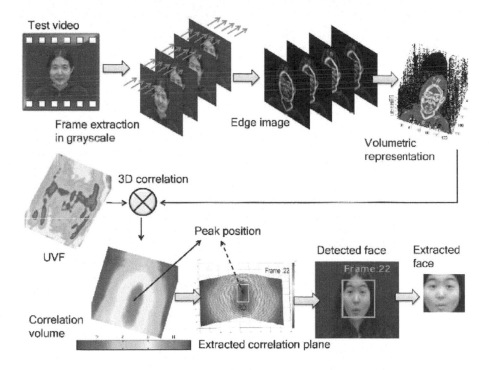

**FIGURE 10.12**: Detailed face detection process from test video using UVF

From the correlation volume correlation planes are extracted framewise. The dimension of each correlation plane is equivalent to $128 \times 128$ as that of the dimension of the target scene. The shift invariant property of UVF makes it easy to locate the point of interest in the target scene and this is reflected in the correlation plane where the maximum value of the response is obtained. This maximum value is called as peak. Each position of peak corresponds to each correlation plane recorded. With respect to the position of peak the region of interest (ROI) is calculated over the correlation plane. As the design of UVF includes only the face part, the ROI is selected on the basis of dimension of the UVF. ROI in one frame is shown in Figure 10.12. Once the ROI is obtained it is now mapped into the original test target scene. The mapping of ROI from correlation plane to target scene is shown in Figure 10.12. It is clear from Figure 10.12 that the ROI in the target scene contains approximately the face part only. Figure 10.12 shows the automatic face detection in video using UVF.

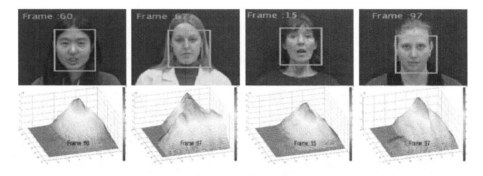

**FIGURE 10.13**: Some detected faces and corresponding correlation planes

Figure 10.13 shows some of the frames of detected faces of different persons in test videos. Figure 10.13 shows the different locations of the distinct peak in the correlation planes corresponding to target frames. The positions of the peak vary according to the movement of face in video.

### 10.10.3.1　Modification in training approach

In previous training section, videos of all persons are taken for synthesizing the UVF. Hence for the VidTIMIT database $43 \times 2 = 86$ videos are needed to train the UVF. As the number of training videos increases, some overfitting problem may occur during the synthesis of UVF which may affect the face detection results, this leads to further modification in training of the UVF. The idea behind face detection is to detect the movement of facial parts in the target scene while a person is talking. With this idea the solution becomes more generalized, i.e. instead of searching of face in a target, the movements of facial parts is searched by the UVF. Hence only the training videos corresponding to one person are used for UVF synthesis and applied over the test videos of another person. Figure 10.14 shows the face detection results of different persons in video while UVF is synthesized by person-1 only.

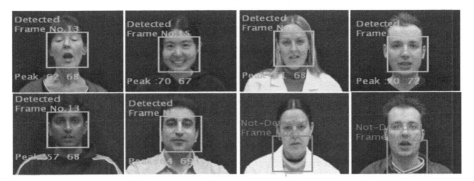

**FIGURE 10.14**: Both detected and not-detected faces are shown for different persons when UVF is synthesized only with person-1

## 10.10.4    Validation of face detection

Validation of face detection can be done by verifying the peak location in the correlation planes (framewise) with the exact location of the nose tip of the face. But during talking as there is some movement of face parts taking place, the location of the nose tip will be somehow displaced. Hence instead of taking the exact location of the nose tip an area ($\pm$10 pixels in both the direction around nose tip corresponding to first frame of test video) around it is considered. This area is recorded manually in beforehand for all persons' test videos. In the testing stage if the peak location resides within this area then only the face is detected. This way the face detection result is validated in this study. Figure 10.14 shows some detected and not-detected faces in some video frames of different persons. A summary of face detection rates for nine random persons is given in Table 10.7 while only person-1 video is used for UVF synthesis. It is interesting to observe in Figure 10.14 that a male face is detected while a female face is used for synthesis and that is due to the fact that the designed UVF searches only the face part movement in video.

## 10.10.5    Face classification using DCCF

Person classification in video is done by testing faces with trained DCCFs. For a test video of size $128 \times 128 \times 135$, 135 faces are extracted as shown in Figure 10.12. Extracted faces are fed to a DCCF bank (contains $L$ number of DCCFs) and $L$ number of correlations are made. According to Equation 10.58, for each extracted test face, $L$ number of distances are obtained. Minimum distance is calculated from $L$ distances. The unknown face is labeled to the class for which the minimum distance is found. Hence in this approach $L + 1$ number of correlations are required where there are $L$ correlations for face classification using DCCF and single 3D correlation for detecting face in video by UVF; whereas in the case of the frame-based target detection approach

**TABLE 10.7**: Face detection rates of different persons in video using UVF synthesized with person-1 are summarized. Average detection rate is also given.

| Person | %Detection | | | | | | | | | %AD |
|---|---|---|---|---|---|---|---|---|---|---|
| | 1 | 2 | 3 | 4 | 5 | 6 | 7 | 8 | 9 | |
| Session1 | 100 | 100 | 100 | 100 | 100 | 100 | 100 | 81.1 | 100 | 97.9 |
| Session2 | 100 | 100 | 92.6 | 100 | 100 | 100 | 100 | 75.4 | 100 | 96.4 |
| Session3 | 100 | 100 | 88.7 | 100 | 100 | 100 | 100 | 0 | 100 | 87.6 |
| Session4 | 100 | 100 | 100 | 100 | 100 | 100 | 100 | 0 | 100 | 88.88 |
| Session5 | 100 | 100 | 100 | 100 | 100 | 100 | 100 | 83.56 | 100 | 98.17 |
| Session6 | 100 | 100 | 100 | 100 | 81.1 | 94 | 100 | 100 | 100 | 97.2 |
| Session7 | 100 | 100 | 100 | 100 | 83 | 100 | 100 | 0 | 100 | 87 |
| Session8 | 100 | 100 | 100 | 100 | 100 | 65.5 | 100 | 91.1 | 100 | 95.2 |
| Session9 | 100 | 100 | 100 | 0 | 100 | 100 | 100 | 100 | 100 | 88.9 |
| Session10 | 100 | 100 | 100 | 0 | 100 | 100 | 100 | 100 | 100 | 88.9 |

using 2D correlation filters, a total of $L + T$ ($T$ = total number of frames in video) number of correlations are required. Obviously ($T >> 1$) and as the length of video increases more time will be needed for detection purposes. Here by using UVF with 3D correlation the detection time is minimized. Figure

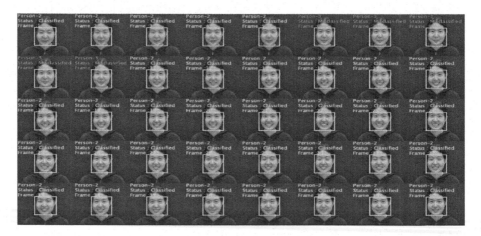

**FIGURE 10.15**: Detection and classification results of person-2 are shown when person-1 is used for UVF training

10.15 shows 40 frames out of 136 frames of Person-2 fourth target video with both detection and classification results. Face detection is validated and then DCCF is used for classification. It has been seen that although the face is detected properly five images are misclassified which is due to DCCF. Five out of 136 frames are misclassified when 40 frames are tested. 96.3% classification accuracy is achieved.

Classification accuracy rate of test person by combining UVF and DCCF algorithm is given in Table 10.8, Table 10.9 and Table 10.10. Table 10.8 provides the accuracy of person classification in session-1 with last the four sentences i.e. overlapping is not considered. Table 10.9 and Table 10.10 gives the classification results of session-2 and session-3, respectively.

**TABLE 10.8**: Classification accuracy rate of 4 persons in video of Session-1 is summarized. TF stands for total frames contained in the respective videos. MF indicates number of frames misclassified

| Session-1 | Sentence-3 | | Sentence-4 | | Sentence-5 | | Sentence-6 | |
|---|---|---|---|---|---|---|---|---|
| | %CR | MF/TF | %CR | MF/TF | %CR | MF/TF | %CR | MF/TF |
| Person-1 | 100 | 0/217 | 100 | 0/72 | 100 | 0/117 | 100 | 0/97 |
| Person-2 | 88.70 | 7/62 | 98.52 | 2/136 | 87.5 | 14/112 | 97.03 | 3/101 |
| Person-3 | 100 | 0/98 | 100 | 0/145 | 100 | 0/81 | 95.34 | 4/86 |
| Person-4 | 100 | 0/139 | 87.75 | 12/98 | 100 | 0/76 | 100 | 0/99 |

**TABLE 10.9**: Classification accuracy rate of 4 persons in video of Session-2 is summarized.

| Session-2 | Sentence-7 | | Sentence-8 | |
|---|---|---|---|---|
| | %CR | MF/TF | %CR | MF/TF |
| Person-1 | 100 | 0/135 | 94.44 | 8/144 |
| Person-2 | 90.14 | 7/71 | 91.26 | 9/103 |
| Person-3 | 100 | 0/97 | 96.63 | 3/89 |
| Person-4 | 100 | 0/71 | 100 | 0/69 |

**TABLE 10.10**: Classification accuracy rate of 4 persons in video of Session-3 is summarized.

| Session-2 | Sentence-9 | | Sentence-10 | |
|---|---|---|---|---|
| | %CR | MF/TF | %CR | MF/TF |
| Person-1 | 100 | 0/126 | 94.5 | 6/109 |
| Person-2 | 88.32 | 16/137 | 91.80 | 10/122 |
| Person-3 | 95.56 | 4/90 | 96.63 | 6/97 |
| Person-4 | 94.44 | 7/126 | 91.3 | 6/69 |

From Table 10.8, Table 10.9 and Table 10.10 the mean %CR for person-2 is observed as 91.66% whereas mean %CR for person-1, person-3 and person-4

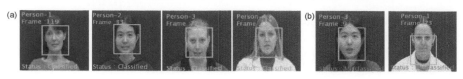

**FIGURE 10.16**: (a) Frames showing accurate classification of different persons, (b) frames showing misclassification occurred during testing

are 98.62%, 98.02% and 96.68% respectively. The minimum %CR is obtained for person-2 comparing to other 3 persons. This is due to the fact there is maximum scale change of face corresponding to person-2 during recitation. Figure 10.16 shows some accurate classification frames as well as misclassified frames obtained during testing.

In Figure 10.16(b) the second image is misclassified due to improper detection of the face part by UVF. The first image in Figure 10.16(b) is misclassified by DCCF due to scale variation while proper face detection (by visual inspection) is made by UVF. Hence results given in Table 10.8, Table 10.9 and Table 10.10 reflect the performance of the combined UVF and DCCF algorithms in automatic face detection and classification.

# Chapter 11

## Subspace-based face recognition in frequency domain

## 11.1 Introduction

Unlike the design of the correlation filter mentioned in the previous chapter, this chapter provides a design method based on subspace-based reconstruction of faces. Class-specific subspace analysis is carried out for the formulation of correlation filters. Face reconstruction using class-specific subspace provides more information in the discriminating stage. Two types of phase-only filters are developed using the projected images and reconstructed images. Correlation between these two filters is used for the classification process.

## 11.2 Subspace-based correlation filter

The correlation filter performances in case of illumination invariant face recognition are reported in [176, 199, 200] where the MACE filter and/or its unconstrained versions are used. In [201] the performance of the phase-only version of the UMACE filter is addressed to handle the illumination variation. In each case, UMACE filter is designed with a set of training images either randomly or systematically chosen from the database so that the designed filter can exhibit precise classification under unknown illumination in test faces. It is not always possible to select the proper training images so that illumination variation of all test faces may lie in the convex hull of training variations. Increasing the number of training images can provide a solution, although it has been reported [202] that in such a case, signal to noise ratio (SNR) will monotonically decrease with the increase in the number of training images.

A solution to this problem may be addressed, if the nature of the correlation filter is changed dynamically according to the input face images so as to achieve robust recognition for all possible illumination variations that lie in a three dimensional (3D) linear subspace for a Lambertian model. Towards achieving this goal, this method is aided by face reconstruction using class-specific subspace analysis. It has been shown in [203] that the low energy in the residue image can be a good criterion to authenticate a face image and it is possible to achieve illumination invariance, if class-specific subspace analysis is performed instead of global subspace analysis. It is shown in Figure 11.1, that the test face image is almost perfectly reconstructed, when the test face is taken from the class of training faces. It is interesting to note from Figure 11.1 when the image of person-3 (taken from PIE database) projected onto the subspace developed by person-1 images, the reconstructed image looks like

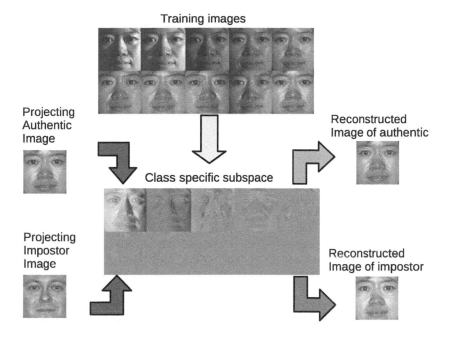

**FIGURE 11.1**: The face reconstruction is shown while the subspace analysis is performed over an individual (person-1) from PIE face.

person-1 after reconstruction. Each individual's orthonormal eigenface basis best spans the face of the same person rather than the other person's face. This can be established by computing the reconstruction error as a squared norm, between the test face and its reconstructed version using Equation 11.1 when projected on someone's (say person-1 here) subspace. Hence all face images from person-1 can be treated as authentic and the other person's face images are impostor images. The error is given by

$$Error = \|t - r\|^2 \tag{11.1}$$

where $t$ is test face and $r$ is its reconstruction.

Hence the observation from Figure 11.1 can be used to discriminate between impostor reconstruction and the authentic reconstruction, where the performance of impostor reconstruction gives higher error. This type of reconstruction helps in discriminating an authentic or an impostor face if a proper filter design is made. Moreover, the phase correlation of filters synthesized from the projecting image and reconstructed image gives a sharp peak for authentic, and no such peak is shown for impostor images. Hence if two filters like $\mathbf{H}_p$ and $\mathbf{H}_r$ can be constructed corresponding to a projecting image and reconstructed image, respectively, then the phase correlation phase correlation between $\mathbf{H}_p$ and $\mathbf{H}_r$ gives a sharp peak in the correlation plane indicating authenticity of the test face, and no such peak in the correlation plane indicates the rejection of the test face as impostor. Phase correlation of two filters is exploited as phase contains more information than magnitude and if the phase spectrum of $\mathbf{H}_p$ and $\mathbf{H}_r$ are cancelled out, a constant flat spectrum is obtained giving a delta-type response in the correlation plane.

In case of the 1DPCA, the one-dimensional data that result from the 2D image by lexicographic ordering create a large covariance matrix and hence the chance of proper analysis with a large number of samples becomes difficult. This problem is overcome by 2DPCA where the computation of covariance matrix involves only 2D data. It has been shown in this study that a subspace-based correlation filter can also be synthesized by 2D subspace analysis. By employing a 2DPCA based class-specific subspace, reconstructed correlation filters are synthesized and a decision can be made as in the case of 1D subspace analysis. The performance of this technique is evaluated on standard illumination face databases like PIE and YaleB.

## 11.3    Mathematical modelling with 1D subspace

### 11.3.1    Reconstructed correlation filter using 1D subspace

Each training face image is of size $(d_1 \times d_2)$. Class-specific subspace analysis is developed over a certain class $C_k$ out of $M$ number of classes $(k = 1, 2, \cdots, M)$ where each class contains $N$ number of lexicographically ordered training vectors $x_i$ of dimension $d \times 1$, where $d = d_1 \times d_2$. It has been suggested in [96] that by withdrawing the first three principal components, the variation due to the lighting condition is reduced. Moreover, the least significant eigenvector is more sensitive to noise [204]. This can be further justified by reconstructing a face from one's individual eigenface subspace and evaluating the reconstruction error as indicated in Equation 11.1. A total of fourteen training images are taken to develop person-1's eigenface subspace and hence fourteen eigenvectors are formed as an orthonormal eigenface basis. One image of person-1 is taken and fourteen reconstructions are performed by taking eigenvectors from one to fourteen and in each case reconstruction error is measured. Error plot is given in Figure 11.2(a) where it is observed that while the first thirteen eigenvectors are taken for reconstruction, minimum error is obtained.

On the other hand, when all fourteen eigenvectors are considered for reconstruction, error is increased. This is clearly visible in Figure 11.2(b), where error is log transformed. As the reconstruction error increases for the inclusion of the least significant eigenvector, it is discarded from the generated subspace in further studies. Hence the truncated subspace for the $k$th class,

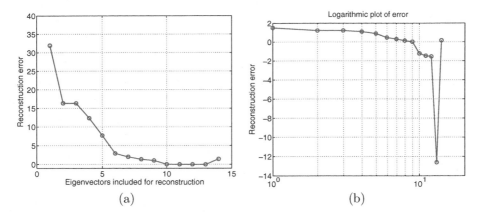

**FIGURE 11.2**: (a) Reconstruction error plot. (b) Error plot in log-error sense. Minimum error is obtained when last eigenvector is excluded during reconstruction

i.e., $E^k$, is obtained by withdrawing first three eigenvectors and hen the last eigenvector and is given as

$$E^k = \begin{bmatrix} c_4 & e_5 & e_6 & \cdots & e_{(N-1)} \end{bmatrix}_{d \times (N-4)} \tag{11.2}$$

where $e_i$'s are the orthonormal vectors.

During reconstruction of face image, for any test image $T^j$ from $j$th class with corresponding vector $\tau^j$ of $d \times 1$ dimension, the difference vector is obtained as

$$\tau_d^j = \tau^j - m \tag{11.3}$$

where $m$ is the average image vector.

Projecting $\tau_d^j$ into the subspace $E^k$, the weight vector $\omega$ is obtained as

$$\omega = [E^k]^{\mathrm{T}} \tau_d^j \tag{11.4}$$

The reconstructed version of the test vector $\tau^j$ can be formulated as

$$r^k = m + \sum_{i=1}^{N-4} e_i^k \omega_i \tag{11.5}$$

The superscript of $r$ in Equation 11.5 is set to $k$ as the reconstruction is done through the orthonormal basis of $k$th class subspace. The reconstructed image $R^k$ (of dimension $d_1 \times d_2$) in the space domain can be reconstructed from Equation 11.5 by reshaping the vector $r^k$ in proper row-column order. The reconstructed correlation filter (RCF), $\mathbf{H}_r^k$, can now be formed by simply taking the Fourier transform of reconstructed image $R^k$ and is given by

$$\mathbf{H}_r^k = \sum_{x=0}^{d_1-1} \sum_{y=0}^{d_2-1} R^k e^{-\frac{j2\pi u x}{d_1}} e^{-\frac{j2\pi v y}{d_2}} \tag{11.6}$$

Though the illumination of the image mostly influences the magnitude spectrum, yet a major benefit is accrued by obtaining the phase spectrum because most images which have more energy at low frequencies, the resulting correlation peaks tend to be sharper than those provided by the matched filter. On the other hand if an image is dominated by high frequencies, the phase filter does not amplify high frequencies as much as an matched filter. As the poorly illuminated images contain more energy at low frequencies, the phase spectrum analysis of these images is a logical choice. The phase spectrum of RCF is given by

$$\mathbf{H}_{r\phi}^k = e^{i\angle\mathbf{H}_r^k}, \qquad i = \sqrt{-1} \tag{11.7}$$

where $\mathbf{H}_{r\phi}^k$ is the phase only RCF corresponding to the reconstructed image obtained by projecting any $j$th class image onto the $k$th class subspace and reconstructed.

### 11.3.2 Optimum projecting image correlation filter using 1D subspace

In addition to RCF another correlation filter is developed simultaneously using the projecting image $T^j$. This correlation filter is designed by minimizing the energy at the correlation plane containing undesired side lobes and maximizing the correlation peak height since a sharp and distinct correlation peak reduces the chances of misclassification. The frequency domain representation of projecting image correlation energy (PICE) at the correlation plane $g^j(m,n)$ in response to input image $T^j$ is given by

$$\text{PICE} = \sum_{u=1}^{d_1}\sum_{v=1}^{d_2}|\mathbf{G}^j(u,v)|^2 \tag{11.8}$$

where $\mathbf{G}^j$ is the Fourier transform of $g^j$.

Let the desired optimum filter be $\mathbf{H}_p^j$. The frequency domain correlation surface $\mathbf{G}^j(u,v)$ is obtained by correlating the Fourier transformed test face image $\mathbf{T}^j$ and the desired filter $\mathbf{H}_p^j$. Hence $\mathbf{G}^j(u,v)$ can be reformulated as

$$
\begin{aligned}
\mathbf{G}^j(u,v) &= \mathbf{H}_p^j(u,v)\mathbf{T}^j(u,v)^* \\
&= \sum_{u=1}^{d_1}\sum_{l=1}^{d_2}|\mathbf{H}_p^j(u,v)|^2|\mathbf{T}^j(u,v)|^2
\end{aligned}
\tag{11.9}
$$

If $\mathbf{H}_p^j$ and $\mathbf{T}^j(u,v)$ are expressed by vector $\mathbf{h}_p^j$ and $\mathbf{t}^j$, then PICE can be expressed by the matrix-vector equation as

$$\text{PICE} = (\mathbf{h}_p^{j+}\bar{\mathbf{T}}^j)(\bar{\mathbf{T}}^{j*}\mathbf{h}_p^j) = \mathbf{h}_p^{j+}\bar{\mathbf{P}}^j\mathbf{h}_p^j \tag{11.10}$$

where $\bar{\mathbf{T}}^j = \mathrm{diag}\{\mathbf{t}^j\}$, and $\bar{\mathbf{P}}^j = \bar{\mathbf{T}}^j\bar{\mathbf{T}}^{j*}$ is a diagonal matrix containing power spectral density of $T^j$ along its diagonal.

In addition to suppressing side lobes of the correlation peak, it is also needed that the required optimum filter must yield large peak values at the origin of the correlation plane. This condition is met by maximizing projecting image correlation height (PICH). The frequency domain expression for PICH for $T^j$ is obtained as,

$$\mathrm{PICH} = \mathbf{t}^{j+}\mathbf{h}_p^j \tag{11.11}$$

To make PICH large while minimizing PICE the optimum filter $\mathbf{h}_p^j$ is synthesized by maximizing the objective function,

$$O(\mathbf{h}_p^j) = \frac{|\mathrm{PICH}|^2}{\mathrm{PICE}} = \frac{\mathbf{h}_p^{j+}\mathbf{t}^j\mathbf{t}^{j+}\mathbf{h}_p^j}{\mathbf{h}_p^{j+}\bar{\mathbf{P}}^j\mathbf{h}_p^j} \tag{11.12}$$

where $|\mathrm{PICH}|^2$ represents the energy of the correlation plane peak value.

Maximizing $O(\mathbf{h}_p^j)$ results in a smaller denominator and hence the term $\bar{\mathbf{P}}^j = \bar{\mathbf{T}}^j\bar{\mathbf{T}}^{j*}$ will be reduced or minimized. Setting the gradient of $O(\mathbf{h}_p^j)$ with respect to $\mathbf{h}_p^j$ to zero then following equation can be formulated,

$$\nabla\{O(\mathbf{h}_p^j)\} = 2\frac{\mathbf{t}^j\mathbf{t}^{j+}\mathbf{h}_p^j}{\mathbf{h}_p^{j+}\bar{\mathbf{P}}^j\mathbf{h}_p^j} - 2\frac{(\mathbf{h}_p^j\mathbf{t}^j\mathbf{t}^{j+}\mathbf{h}_p^j)(\bar{\mathbf{P}}^j\mathbf{h}_p^j)}{\{\mathbf{h}_p^{j+}\bar{\mathbf{P}}^j\mathbf{h}_p^j\}^2} = 0 \tag{11.13}$$

Considering,

$$\lambda = \frac{\mathbf{h}_p^{j+}\mathbf{t}^j\mathbf{t}^{j+}\mathbf{h}_p^j}{\mathbf{h}_p^{j+}\bar{\mathbf{P}}^j\mathbf{h}_p^j} \tag{11.14}$$

Equation 11.13, reduces to

$$\frac{1}{\mathbf{h}_p^{j+}\bar{\mathbf{P}}^j\mathbf{h}_p^j}\{\mathbf{t}^j\mathbf{t}^{j+}\mathbf{h}_p^j - \lambda\bar{\mathbf{P}}^j\mathbf{h}_p^j\} = 0 \tag{11.15}$$

From Equation 11.15 the following equation can be written as,

$$\mathbf{t}^j\mathbf{t}^{j+}\mathbf{h}_p^j - \lambda\bar{\mathbf{P}}^j\mathbf{h}_p^j = 0 \tag{11.16}$$

$\bar{\mathbf{P}}^j$ is a diagonal matrix and the inversion of this is trivial. As $\bar{\mathbf{P}}^j$ is invertible Equation 11.16 can be written as

$$\lambda\mathbf{h}_p^j = [\bar{\mathbf{P}}^j]^{-1}\mathbf{t}^j\mathbf{t}^{j+}\mathbf{h}_p^j \tag{11.17}$$

It is evident that Equation 11.12 and Equation 11.14 are identical. As the product $\mathbf{t}^j\mathbf{t}^{j+}$ has unit rank, the term $[\bar{\mathbf{P}}^j]^{-1}\mathbf{t}^j\mathbf{t}^{j+}$ has only one non-zero eigenvalue and the corresponding eigenvector maximizes the objective function $O(\mathbf{h}_p^j)$. Hence the optimum filter $\mathbf{h}_p^j$ is the eigenvector corresponding to the eigenvalue $\lambda$. Without loss of generality it may be assumed that the value at the origin of correlation plane is a scalar quantity and is given as

$$\mathbf{t}^{j+}\mathbf{h}_p^j = \beta \tag{11.18}$$

Hence from Equation 11.18 and Equation 11.17 the following equation can be written as

$$\beta[\bar{\mathbf{P}}^j]^{-1}\mathbf{t}^j = \lambda\mathbf{h}_p^j \tag{11.19}$$

or

$$\mathbf{h}_p^j = \frac{\beta}{\lambda}\{[\bar{\mathbf{P}}^j]^{-1}\mathbf{t}^j\} \tag{11.20}$$

where $\frac{\beta}{\lambda}$ is a scalar quantity and acts as a scale factor.

The closed form solution of $\mathbf{h}_p^j$ in Equation 11.20 represents the optimum single image unconstrained filter vector. The 2D-filter $\mathbf{H}_p^j$ corresponding to $\mathbf{h}_p^j$ can be obtained by proper row-column arrangement. For better representation of peak sharpness, the phase spectrum of the filter is used. The phase representation of the 2D optimum filter is denoted by

$$\mathbf{H}_{p\phi}^j = e^{i\angle\mathbf{H}_p^j}, \qquad i = \sqrt{-1} \tag{11.21}$$

## 11.4 Face classification and recognition analysis in frequency domain

The detailed process of face recognition is given in Figure 11.3. At first, the class-specific subspace is computed over the total population of $M$ classes of face images. Hence $M$ numbers of class-specific subspaces $E^k, (k = 1, 2, \cdots, M)$ are formed. A test face image $T^j$ (treated as a projecting image) from any $j$th class $(j \in M)$ is projected onto the $M$ numbers of subspaces resulting in $M$-numbers of reconstructed images $R^k$. From these $M$ number of $R^k$, the reconstructed correlation filter (RCF) for each class is formed according to Equation 11.7. Hence $M$ number of $\mathbf{H}_{r\phi}^k$ are formed. Along with this operation, the phase-only OPICF $\mathbf{H}_{p\phi}^j$ is formed from test face image $T^j$.

The spatial domain correlation output obtained from $\mathbf{H}_{r\phi}^k$ and $\mathbf{H}_{p\phi}^j$ is given by

$$g(m,n) = FFT^{-1}\{\mathbf{H}_{r\phi}^k \otimes \mathbf{H}_{p\phi}^{j*}\} \tag{11.22}$$

Decision of authentication can be made in an ideal situation by using the following relation,

$$
\begin{aligned}
g(m,n) &= \delta(m,n), \quad \text{if} \quad j = k \\
&= \text{random matrix}, \quad \text{otherwise}
\end{aligned} \tag{11.23}
$$

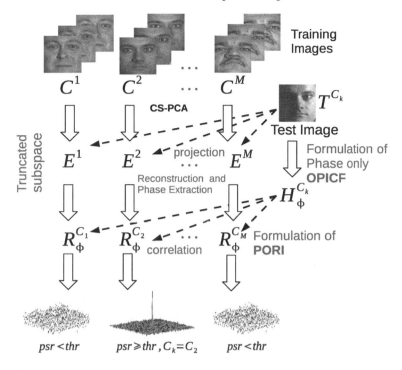

**FIGURE 11.3**: Detailed process of the present system

From the correlation planes PSRs are calculated. The test face image is classified into a class for which the PSR value is greater than the preset-set threshold value denoted by *thr* (shown in Figure 11.3), usually taken as 10 [161].

## 11.5 Test results with 1D subspace analysis

### 11.5.1 Comparative study in terms of PSRs

Initially a comparative study is made with the help of correlation planes from where the PSR values are calculated for making a decision. All the standard filters are trained with the same set of training images from the YaleB database and tested over a non-trained authentic image. Correlation planes corresponding to all filters are shown in Figure 11.4. Here images of subset-3 are used for training and the first image from subset-5 is used for testing. PSRs are calculated from each correlation plane. It is observed from

Figure 11.4 that the highest PSR is obtained for this system. If the threshold value of PSR is set to 10 for authentication then it can be claimed that the present system truly accepts the authentic image whereas the other filters reject the true images.

The same test is performed for PIE database where image index $10, 19$ are chosen for training as the two images are frontal lighting face images. The first image from the PIE database is taken for test and the correlation planes in response to this are shown in Figure 11.5. It is again observed from Figure 11.5 that the highest PSR ($> 10$) is obtained in the case of this type of filtering approach. Further, the test result is extended for observing the PSR values

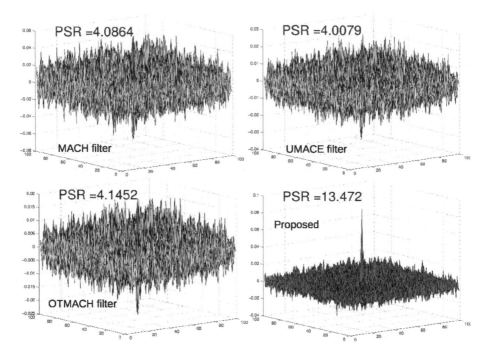

**FIGURE 11.4**: Correlation planes in response to the first image from subset-5 for different filters. In this case the correlation plane shows a sharp and distinct peak for authentic comparison to others

over the whole YaleB database. In this case all the filters are synthesized with images from subset-1 to subset-4. Figure 11.6 shows the comparative performance.

## 11.5.2    Comparative study on %RR and %FAR

It is necessary to measure %FAR as well as % RR as an indication for robust face recognition. High PSR confirms that the recognition result would be better, but simultaneously it may so happen that in the case of impostors,

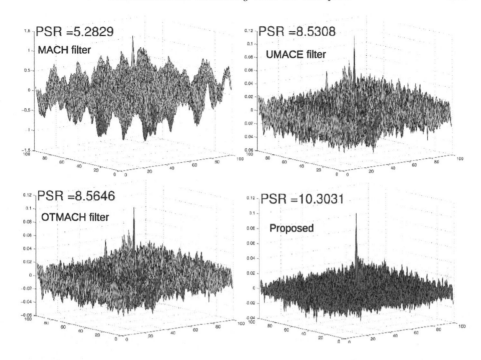

**FIGURE 11.5**: Correlation planes in response to the first image from PIE for different filters. In this case the correlation plane shows better PSR value in response to authentic

the PSR values may also be high, resulting in an increase in FAR. Hence to observe the performance of the present system, it is necessary to record %FAR along with %RR. Based on the PIE and YaleB database, the comparative performance is given in Table 11.1. Three sets of training images are taken randomly from both PIE and Yale databases to synthesize the filters. Table 11.1 summarizes the results obtained after performing the experiments over the whole database. The mean recognition and mean error rate are calculated and presented in Table 11.1.

## 11.6    Mathematical modelling with 2D subspace

For the completeness of this study, 2DPCA is carried out for class-specific subspace analysis. 2D class-specific subspace is employed for the reconstruction of face images from which a pair of correlation filters, i.e., RCF and OPICF are formed. The method of making a decision is evolved by

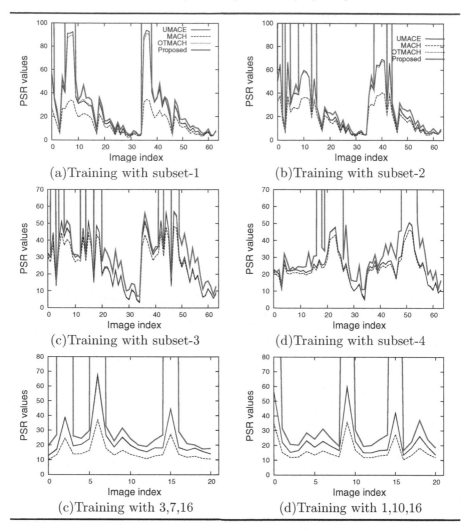

**FIGURE 11.6**: PSR distributions of faces from YaleB database for all filters

phase correlation of OPICF and RCF as discussed earlier. Here also the nature of RCF and OPICF is dynamically changed from image to image. Again if the input image is from any class with which the subspace is developed, the correlation between RCF and OPICF gives a distinct peak in the correlation plane from which the authentication of the input face is tested.

## 11.6.1 Reconstructed correlation filter using 2D subspace

If $A_{ik}$ is the $i^{th}$ training image of size $d_1 \times d_2$ (where $i = 1, \cdots, N$), of $k^{th}$ class, (where $k = 1, \cdots, M$), and $N$ is the total number of training images in

**TABLE 11.1**: Summary of %mean recognition and %mean error rate comparison of all filters while experiments are performed with both the databases.

| Methods | Yale Train-1 | | Yale Train-2 | | Yale Train-3 | |
|---|---|---|---|---|---|---|
| | % rec | % error | % rec | % error | % rec | % error |
| UMACE | 89.06 | 3.716 | 96.87 | 0.42 | 95.3125 | 5.405 |
| UOTSDF | 87.5 | 6.33 | 96.87 | 1.82 | 96.87 | 7.51 |
| MACH | 90.62 | 5.02 | 96.87 | 1.65 | 96.8 | 6.37 |
| OTMACH | 89.06 | 3.97 | 96.87 | 0.55 | 95.317 | 5.82 |
| Proposed | 96.87 | 0.211 | 100 | 0.38 | 100 | 0.2534 |
| Methods | PIE Train-1 | | PIE Train-2 | | PIE Train-3 | |
| | % rec | % error | % rec | % error | % rec | % error |
| UMACE | 100 | 1.2698 | 100 | 1.507 | 100 | 0.9524 |
| UOTSDF | 100 | 1.2698 | 100 | 1.5079 | 100 | 0.9524 |
| MACH | 100 | 1.4286 | 100 | 1.9048 | 100 | 1.1905 |
| OTMACH | 100 | 1.4286 | 100 | 1.9841 | 100 | 1.1905 |
| Proposed | 100 | 0.92 | 100 | 1.03 | 100 | 0.1587 |

one class and $M$ is the total number of classes in a given database, then the average of the training face images of the $k^{th}$ class is given by

$$\bar{A}_k = \frac{1}{N} \sum_{i=1}^{N} A_{ik} \qquad (11.24)$$

The optimal projection axes $x_1, \cdots, x_{d_2}$ are the $d_2$ number of orthonormal eigenvectors of image covariance matrix of size $d_2 \times d_2$. Having obtained orthonormal eigenvectors, a matrix $U_k$ is formed by placing the eigenvectors as,

$$U_k = \begin{bmatrix} x_1 & x_2 & \cdots & x_{d_2} \end{bmatrix} \qquad (11.25)$$

where, $U_k$ is the class-specific eigen matrix.

For reconstruction purposes any test image $T^j$ of $j, (j = 1, \cdots, M)$th class is projected on the eigen matrix $U_k$ according to the following equation given by

$$V = T^j U_k \qquad (11.26)$$

where $V$ is the principal component matrix.

The representative of a certain class plays an important role for class-specific reconstruction. Without loss of generality, the mean image $\bar{A}_k$ is considered as the representative of the $k$th class. Hence the class-specific reconstruction is obtained as,

$$R^k = \bar{A}_k + V(U_k)^{\mathrm{T}} \qquad (11.27)$$

where, $R^k$ is the reconstructed image of the $T^j$ input image when projected on the $k$th class subspace.

The 2D-Fourier transform of reconstructed image $R^k$ is given by

$$\mathbf{H}_r^k = \sum_{x=0}^{d_1-1} \sum_{y=0}^{d_2-1} R^k(x,y) e^{-\frac{j2\pi ux}{d_1}} e^{-\frac{j2\pi vy}{d_2}} \tag{11.28}$$

where, $\mathbf{H}_r^k$ is the frequency domain counterpart of $R^k$. Hence the phase spectrum of the RCF is given by

$$\mathbf{H}_{r\phi}^k = e^{i\angle \mathbf{H}_r^k} \qquad i = \sqrt{-1} \tag{11.29}$$

In addition to the reconstructed correlation filter (RCF) another correlation filter OPICF is formed. The synthesis of OPICF does not involve the subspace analysis. Hence the designed OPICF and its phase-only version can be used as obtained in Equation 11.21. Face classification and recognition technique is the same as described in Sec.11.4 with one difference: 2D subspace analysis is performed instead of 1D subspace.

---

## 11.7 Test results on 2D subspace analysis

### 11.7.1 PSR value distribution for authentic and impostor classes

The class-specific 2DPCA is performed for each person from PIE (no light dataset) using three images with extreme lighting variations, i.e. the image-1(left shadow), image-10(frontal lighting) and image-16(right shadow). Each image is projected and reconstructed from where RCF and OPICF are formed. Two phase spectra as obtained are correlated to get the response surface and the PSR values are recorded. It is observed that when the the test images are selected from the classes for which the 2DPCA is developed, the corresponding PSR values are high and for other class images the PSRs are low. Figure 11.7 is obtained by synthesizing 2D subspace with training face images (1,10,16) of person-1 and all $65 \times 21$ face images are tested. Twenty-one maximum PSR values are recorded from impostor classes and PSR values of person-1 images are plotted. From Figure 11.7 it may be said that for both identification and verification an accuracy of 100% is achieved.

### 11.7.2 Comparative performance in terms of %RR

To evaluate the performance of this filter other tests are performed on the illumination subset of PIE and YaleB face databases. These databases exhibit large illumination variation and there are many face images with substantial shadow. Table 11.2 and Table 11.3 shows the %mean recognition rate of the

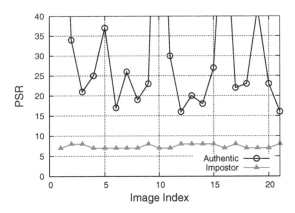

**FIGURE 11.7**: PSR values for both authentic class and impostor class images. Here the authentic class is person-1

PIE database while the images are captured from extreme lighting variations and near frontal lighting, respectively. It is observed from the Table 11.2 that

**TABLE 11.2**: Observation of top %mean recognition comparison of different frequency domain correlation filter methods while all the filters are tested over whole database (PIE with room light off).

| | Training Images (extreme lighting variations) | | | | |
|---|---|---|---|---|---|
| Methods | 3,7,16 | 1,10,16 | 2,7,16 | 4,7,13 | 3,10,16 |
| MACH | 98.67% | 99.68% | 99.29% | 97.26% | 99.92% |
| OTMACH | 100% | 100% | 100% | 99.92% | 100% |
| UMACE | 100% | 100% | 100% | 99.85% | 100% |
| QPUMACE | 100% | 100% | 100% | 98.75% | 99.92% |
| Proposed | 100% | 100% | 100% | 100% | 100% |

this strategy gives 100% verification results over the whole database.

Similar comparative observations of %mean recognition rates are made for the YaleB database. In this test different subsets are used for training purposes. The performance of different methods is recorded by testing over the whole database (including the training images, i.e. overlapping is considered). Table 11.4 shows the improvement in recognition rate when the present method is employed. It is interesting to note that a large improvement in recognition rate occurred compared to other methods while Subset-1 is used for training, where only a few images are selected with near frontal lighting.

**TABLE 11.3**: Observation of verification accuracy in terms of %mean Recognition while all the filters are tested over whole database (PIE with room light off). Near frontal lighting images are used for training. The $PSR \geq 10$ is selected as pre-set threshold.

| Methods | Training Images (near frontal lighting) | | | | | |
|---|---|---|---|---|---|---|
| | 5,6,7,8,9,10, 11,18,19,20 | 8,9,10 | 18,19,20 | 7,10,19 | 5,7,9,10 | 5,6,7,8, 9,10 |
| MACH | 100% | 97.34% | 98.43% | 71.8% | 95% | 99.34% |
| OTMACH | 99.53% | 99.22% | 99.61% | 96.25% | 95.78% | 98.83% |
| UMACE | 99.76% | 99.22% | 99.45% | 96.17% | 95.8% | 98.6% |
| QPUMACE | 98.75% | 97.97% | 97.58% | 97.35% | 93.6% | 96.17% |
| Proposed | 100% | 99.61% | 99.69% | 99.37 % | 98.51% | 99.61% |

Hence the present system has the ability to recognize true class face images even if the unknown face images are from different illumination conditions.

**TABLE 11.4**: Recognition rate comparison of different filters while the tests are performed over the whole database of YaleB. Different subsets are used for training purposes.

| Methods | Subset-1 | Subset-2 | Subset-3 | Subset-4 | Subset-5 |
|---|---|---|---|---|---|
| MACH | 74.22% | 83.9% | 94.37% | 97.18% | 98.43% |
| OTMACH | 84.37% | 86.40% | 94.53% | 97.18% | 98.59% |
| UMACE | 84.84% | 86.72% | 94.37% | 97.34% | 98.43% |
| QPUMACE | 73.59% | 72.5% | 86.56% | 85.93% | 84.68% |
| Proposed | 94.37% | 92.18% | 98.28% | 99.53% | 99.68% |

### 11.7.3   Performance evaluation using ROC analysis

The performance of correlation filters can be characterized in terms of ROCs. ROCs are calculated for the YaleB database with increasing PSRs as thresholds. When comparing ROC curves of different tests, desirable curves lie closer to the top left corner. It is observed from Figure 11.8, ROC curves corresponding to the subspace-based correlation filter are approaching to a step function and hence it has the best possible detection performance compared to other filters. Advantages of subspace-dependent correlation scheme are clearly visible from the convex hull nature of ROCs in Figure 11.8, where high recognition rates are consistently obtained while FAR is reduced. It is observed that class-dependent subspace-based correlation filtering has

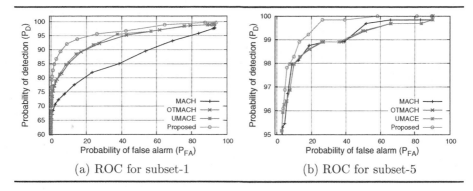

(a) ROC for subset-1          (b) ROC for subset-5

**FIGURE 11.8**: ROC curves for different face classifiers in frequency domain for YaleB

almost 99% verification accuracy with remarkable improvement compared to others. Hence the newly designed system exhibits built-in tolerance to illumination variations.

## 11.8   Class-specific nonlinear correlation filter

. There are other problems in using linear correlation filters range and, hence, in the testing stage it is hard to discriminate authentic and impostor images that lie below a span of low gray levels. To overcome this situation a nonlinear correlation filter can be exploited by using the point nonlinearities [205] of image pixels so that the designed correlation filter achieves a uniform dynamic range. This type of nonlinear mapping stretches pixel distribution of face images in a wide range and consequently high frequency components are amplified.

In [206] three approaches are judicially combined to improve face recognition results under illumination variation, viz., i) projection-based method of designing a correlation filter is used to improve upon the capability of recognition at all possible illumination variations; ii) phase correlation method is used to enhance peak sharpness at the correlation plane for authentic face image as phase contains more information than the magnitude of spectrum and iii) point nonlinearities are considered to extend the uniform dynamic range. To achieve these, two correlation filters are designed: (a) nonlinear optimum projecting image correlation filter $\mathbf{H}_p$ and (b) nonlinear optimum reconstructed image correlation filter $\mathbf{H}_r$. The nature of the design process of these two filters is the same with the only difference in image used for synthesis. Design of $\mathbf{H}_p$ uses a projecting image and the design of $\mathbf{H}_r$ includes a reconstructed image. The phase correlation between these two filters produces a response surface, the nature of which totally depends on the face class involved. Ideally a delta-type peak at the correlation plane is obtained if these two filters are generated from the same face class. Experimental results on standard databases like YaleB (extended) [207] and PIE database [208] witness the promising performance of the new method when compared to other standard correlation filter-based face recognition systems.

## 11.9   Formulation of nonlinear correlation filters

### 11.9.1   Nonlinear optimum projecting image correlation filter

Any projecting image from the $k$th (where, $k = 1, 2, \cdots, M$) class is represented in a spatial domain as $T$ (in matrix form) or $t$ (in vector form) and its frequency domain counterparts are $\mathbf{T}$ and $\mathbf{t}$, respectively. $\bar{\mathbf{T}}$ represents the diagonal form of $\mathbf{t}$. The pointwise nonlinearities of an image can be achieved according to power law transformation given by

$$t_\alpha^\beta = \alpha[t]^\beta \tag{11.30}$$

where $\alpha > 0$ can take any integer value and $\beta > 0$ can be an integer or fraction.

Equation 11.30 shows that each element of $t$ is scaled by $\alpha$th amount and raised to $\beta$th power. Hence for $\alpha = \beta = 1$, the image $t_1^1$ represents the original image $t$. If $\mathbf{h}_{\alpha\beta}$ is the optimum correlation filter corresponding to the projecting image $\mathbf{t}_\alpha^\beta$, then the correlation plane $\mathbf{g}_{\alpha\beta}$ in response to $\mathbf{t}_\alpha^\beta$ is given by

$$\mathbf{g}_{\alpha\beta} = \bar{\mathbf{T}}_\alpha^{\beta*} \mathbf{h}_{\alpha\beta} \tag{11.31}$$

where * represents conjugation operation.

From Equation 11.31, it can be noted that a number of correlation planes $\mathbf{g}_{\alpha\beta}$ as well as a number of classifiers $\mathbf{h}_{\alpha\beta}$ are generated for each value of $\alpha = \alpha_1, \cdots, \alpha_n$ and $\beta = \beta_1, \cdots, \beta_m$ in response to a single projecting image. The same can be written as
$$\bar{\mathbf{T}}_{\alpha_1}^{\beta_1}, \bar{\mathbf{T}}_{\alpha_1}^{\beta_2} \cdots, \bar{\mathbf{T}}_{\alpha_1}^{\beta_m}, \bar{\mathbf{T}}_{\alpha_2}^{\beta_1}, \bar{\mathbf{T}}_{\alpha_2}^{\beta_2}, \cdots, \bar{\mathbf{T}}_{\alpha_2}^{\beta_m}, \cdots, \bar{\mathbf{T}}_{\alpha_n}^{\beta_1}, \bar{\mathbf{T}}_{\alpha_n}^{\beta_2}, \cdots, \bar{\mathbf{T}}_{\alpha_n}^{\beta_m}$$
Hence from Equation 11.31 a set of correlation planes can be written as,

$$\begin{aligned}
\mathbf{g}_{\alpha_1\beta_1} &= \bar{\mathbf{T}}_{\alpha_1}^{\beta_1*} \mathbf{h}_{\alpha_1\beta_1} \\
\mathbf{g}_{\alpha_1\beta_2} &= \bar{\mathbf{T}}_{\alpha_1}^{\beta_2*} \mathbf{h}_{\alpha_1\beta_2} \\
\vdots &= \vdots \\
\mathbf{g}_{\alpha_n\beta_m} &= \bar{\mathbf{T}}_{\alpha_n}^{\beta_m*} \mathbf{h}_{\alpha_n\beta_m}
\end{aligned} \tag{11.32}$$

Since a sharp and distinct correlation peak in the correlation plane reduces the chances of misclassification, minimization of energy at the correlation plane [137] containing undesired side lobes and maximization of correlation peak height[140] are necessary. These criteria help in amplifying the high frequency components of the projecting image of which the point wise nonlinear transformation is done. Hence for selected variations of $\alpha$ and $\beta$, correlation plane energy could be evaluated for each correlation plane indicated in Equation 11.32 as,

$$
\begin{aligned}
|\mathbf{g}_{\alpha_1\beta_1}|^2 &= |\bar{\mathbf{T}}_{\alpha_1}^{\beta_1*}\mathbf{h}_{\alpha_1\beta_1}|^2 = \mathbf{h}_{\alpha_1\beta_1}^{+}\bar{\mathbf{T}}_{\alpha_1}^{\beta_1}\bar{\mathbf{T}}_{\alpha_1}^{\beta_1*}\mathbf{h}_{\alpha_1\beta_1} \\
|\mathbf{g}_{\alpha_1\beta_2}|^2 &= |\bar{\mathbf{T}}_{\alpha_1}^{\beta_2*}\mathbf{h}_{\alpha_1\beta_2}|^2 = \mathbf{h}_{\alpha_1\beta_2}^{+}\bar{\mathbf{T}}_{\alpha_1}^{\beta_2}\bar{\mathbf{T}}_{\alpha_1}^{\beta_2*}\mathbf{h}_{\alpha_1\beta_2} \\
\vdots &= \vdots = \vdots \\
|\mathbf{g}_{\alpha_n\beta_m}|^2 &= |\bar{\mathbf{T}}_{\alpha_n}^{\beta_m*}\mathbf{h}_{\alpha_n\beta_m}|^2 = \mathbf{h}_{\alpha_n\beta_m}^{+}\bar{\mathbf{T}}_{\alpha_n}^{\beta_m}\bar{\mathbf{T}}_{\alpha_n}^{\beta_m*}\mathbf{h}_{\alpha_n\beta_m}
\end{aligned}
\tag{11.33}
$$

Hence to get a sharp peak in each correlation plane for selected variations of $\alpha$ and $\beta$, it is needed to minimize the correlation energies separately, with respect to $\mathbf{h}$, given in Equation 11.33. Minimization of the correlation plane energy is reflected by the following performance criteria of the desired filter $\mathbf{h}_{\alpha\beta}$ and is given by

$$
min\{\mathbf{h}_{\alpha\beta}^{+}\mathbf{T}_{\alpha}^{\beta}\mathbf{T}_{\alpha}^{\beta*}\mathbf{h}_{\alpha\beta}\}
\tag{11.34}
$$

Minimization of the performance criteria, indicated in Equation 11.34, is evaluated with respect to $\mathbf{h}$. It can be noted that the expression in Equation 11.34 is different from the standard performance criteria of the MACE filter, since a set of classifiers has been taken into consideration using point nonlinearities in addition to different scaled magnitudes of image pixels. Now origin value or the peak value of the correlation plane in response to the projecting image $t_{\alpha}^{\beta}$ can be formulated in the frequency domain as $t_{\alpha}^{\beta+}\mathbf{h}_{\alpha\beta}$. In addition to suppressing side lobes of the correlation peak, which can be achieved by Equation 11.34, it is necessary for an optimum filter to yield a large peak value at the origin of the correlation plane. This condition is met by maximizing the projecting image correlation peak intensity with respect to $\mathbf{h}$ for a typical set of $\alpha, \beta$, as,

$$
max\{|t_{\alpha}^{\beta+}\mathbf{h}_{\alpha\beta}|^2\}
\tag{11.35}
$$

Hence, to get optimum correlation filter $\mathbf{h}_{\alpha\beta}$, the optimal trade-off performance criterion can now be set as

$$
J(\mathbf{h}_{\alpha\beta}) = \frac{|t_{\alpha}^{\beta+}\mathbf{h}_{\alpha\beta}|^2}{\mathbf{h}_{\alpha\beta}^{+}\bar{\mathbf{T}}_{\alpha}^{\beta}\bar{\mathbf{T}}_{\alpha}^{\beta*}\mathbf{h}_{\alpha\beta}}
\tag{11.36}
$$

The criterion can be obtained as the dominant eigenvector [140] of $\{\bar{\mathbf{T}}_{\alpha}^{\beta}\bar{\mathbf{T}}_{\alpha}^{\beta*}\}^{-1}t_{\alpha}^{\beta}t_{\alpha}^{\beta+}$. The desired filter is therefore given by

$$
\mathbf{h}_{\alpha\beta} = \{\bar{\mathbf{T}}_{\alpha}^{\beta}\bar{\mathbf{T}}_{\alpha}^{\beta*}\}^{-1}t_{\alpha}^{\beta}
\tag{11.37}
$$

For different values of $\alpha$ and $\beta$, Equation 11.37 can be expanded and expressed in a closed form solution as

$$
\begin{pmatrix} \mathbf{h}_{\alpha_1\beta_1} \\ \mathbf{h}_{\alpha_1\beta_2} \\ \vdots \\ \mathbf{h}_{\alpha_n\beta_m} \end{pmatrix} = \begin{pmatrix} \bar{\mathbf{P}}_{\alpha_1\beta_1} & 0 & \cdots & 0 \\ 0 & \bar{\mathbf{P}}_{\alpha_1\beta_2} & \cdots & 0 \\ \vdots & \vdots & \ddots & 0 \\ 0 & 0 & \cdots & \bar{\mathbf{P}}_{\alpha_n\beta_m} \end{pmatrix}^{-1} \begin{pmatrix} t_{\alpha_1}^{\beta_1} \\ t_{\alpha_1}^{\beta_2} \\ \vdots \\ t_{\alpha_n}^{\beta_m} \end{pmatrix} \tag{11.38}
$$

where $\bar{\mathbf{P}}_{\alpha\beta} = \bar{\mathbf{T}}_\alpha^\beta \bar{\mathbf{T}}_\alpha^{\beta*}$.

Denoting block vectors and matrix as

$$
[\mathbf{h}_p^k] = \begin{pmatrix} \mathbf{h}_{\alpha_1\beta_1} \\ \vdots \\ \mathbf{h}_{\alpha_n\beta_m} \end{pmatrix}, \quad [\bar{\mathbf{P}}] = \begin{pmatrix} \bar{\mathbf{P}}_{\alpha_1\beta_1} & 0 & \cdots & 0 \\ 0 & \bar{\mathbf{P}}_{\alpha_1\beta_2} & \cdots & 0 \\ \vdots & \vdots & \ddots & 0 \\ 0 & 0 & \cdots & \bar{\mathbf{P}}_{\alpha_n\beta_m} \end{pmatrix}, \quad [\mathbf{t}] = \begin{pmatrix} t_{\alpha_1\beta_1} \\ \vdots \\ t_{\alpha_n\beta_m} \end{pmatrix}
$$

Equation 11.38 can be redrawn as

$$
[\mathbf{h}_p^k] = [\bar{\mathbf{P}}]^{-1}[\mathbf{t}] \tag{11.39}
$$

where $[\mathbf{h}_p^k]$ correspond to projecting image from the $k$th class.

As $[\bar{\mathbf{P}}]$ is a block matrix in diagonal form, the decoupled nature of $\mathbf{h}_{\alpha\beta}$ is accomplished, i.e. $\mathbf{h}_{\alpha_1\beta_1}$ depends only on $t_{\alpha_1}^{\beta_1}$ and no other variations of $\alpha$ and $\beta$ are allowed. It can also be stated that $\mathbf{h}_{\alpha_i\beta_l}$ ($0 < i \leq n, 0 < l \leq m$) depends on the nonlinear characteristic of image pixels in $t_\alpha^\beta$ with respect to the original pixels' distribution in $t$ and $\mathbf{h}_{\alpha_i\beta_l}$ is optimally designed with the performance criteria given in Equation 11.34 and Equation 11.35. In addition to that, $\mathbf{h}_{\alpha_i\beta_l}$ is generated from the projecting image. Hence the designed filter given in Equation 11.39 can be termed as a nonlinear optimum projecting image correlation filter or NOPICF.

The 2D correlation filter $\mathbf{H}_{\alpha_i\beta_l}$ ($0 < i \leq n, 0 < l \leq m$) is obtained by reshaping the filter vector $\mathbf{h}_{\alpha_i\beta_l}$ in proper row-column order. Hence by block matrix form 2D-NOPICF for the $k$th class projecting image is expressed as

$$
[\mathbf{H}_p^k] = \begin{pmatrix} \mathbf{H}_{\alpha_1\beta_1} \\ \mathbf{H}_{\alpha_1\beta_2} \\ \vdots \\ \mathbf{H}_{\alpha_n\beta_m} \end{pmatrix}_p \tag{11.40}
$$

where suffix $p$ represents that the filters are synthesized with projecting image.

Equation 11.40 also indicates that the desired 2D-NOPICF $[\mathbf{H}_p^k]$ is a collection of nonlinear classifiers.

## 11.9.2 Nonlinear optimum reconstructed image correlation filter

To reconstruct the projecting image in the nonlinear optimum reconstructed image correlation filter, it is necessary to develop the class-specific subspace. Hence the subspace analysis is made over the $j$th class $(j = 1, 2, \cdots, M)$ where each class contains $N$ number of lexicographic ordered training vectors $x_i$ of dimension $d \times 1$. As the least significant eigenvector is sensitive to noise [204] and may give error during reconstruction, this is discarded from the generated subspace. Therefore, the truncated subspace $E^j$ is formed as,

$$E^j = [e_1, e_2, e_3, \cdots, e_{(N-1)}]_{d \times (N-1)} \tag{11.41}$$

where $e_i$s are the orthonormal vectors and superscript $j$ indicates that the subspace is originated from $j$th class training images.

Since the projecting image can be from any class, the $k$th class test image is considered as the projecting image. During reconstruction of face images, the difference vector of non linearly mapped projecting vector $t_{\alpha_i}^{\beta_l}$ ($0 < i \leq n$, $0 < l \leq m$) is obtained as,

$$s_{\alpha_i}^{\beta_l} = t_{\alpha_i}^{\beta_l} - m \tag{11.42}$$

where $m$ is the average image vector of original training variations $(x_i)$.

Projecting $s_{\alpha_i}^{\beta_l}$ into the subspace $E^j$, the weight vector $\omega_{\alpha_i}^{\beta_l}$ is obtained as

$$\omega_{\alpha_i}^{\beta_l} = (E^j)^{\mathrm{T}} s_{\alpha_i}^{\beta_l} \tag{11.43}$$

where T represents transpose operation.

The reconstructed version $r_{\alpha_i}^{\beta_l}$ corresponding to the test vector $t_{\alpha_i}^{\beta_l}$ is obtained as

$$r_{\alpha_i}^{\beta_l} = m + \sum_{i=1}^{N-1} e_i^j \omega_{\alpha_i}^{\beta_l} \tag{11.44}$$

For different values of $\alpha$ and $\beta$, a set of reconstructed vectors are formed. It is easier to represent these vectors in block vector form as

$$[r^{jk}] = \begin{pmatrix} r_{\alpha_1}^{\beta_1} \\ r_{\alpha_1}^{\beta_2} \\ \vdots \\ r_{\alpha_n}^{\beta_m} \end{pmatrix} \tag{11.45}$$

The superscript $jk$ in Equation 11.45 represents the reconstructed vectors corresponding to the $k$th class test image while projected on the $j$th class subspace. The reconstructed image $R_\alpha^\beta$ in space domain can be obtained by reshaping the vector $r_\alpha^\beta$ in proper row-column order. The frequency domain transformation of the reconstructed image $R_\alpha^\beta$ is simply obtained as,

$$\mathbf{R}_\alpha^\beta = \sum_{p=0}^{d_1-1} \sum_{q=0}^{d_2-1} R_\alpha^\beta(p, q) e^{-\frac{j2\pi up}{d_1}} e^{-\frac{j2\pi vq}{d_2}} \tag{11.46}$$

From Equation 11.46 the NORICF is formed in the same way as NOPICF is designed. Instead of $\mathbf{T}_\alpha^\beta$, however, $\mathbf{R}_\alpha^\beta$ is used for NORICF design. A number of NORICFs are formed for different values of $\alpha$ and $\beta$ as obtained in the case of NOPICFs. Hence with the help of Equation 11.40 the block matrix form of 2D-NORICF is written as,

$$[\mathbf{H}_r^{jk}] = \begin{pmatrix} \mathbf{H}_{\alpha_1\beta_1} \\ \mathbf{H}_{\alpha_1\beta_2} \\ \vdots \\ \mathbf{H}_{\alpha_n\beta_m} \end{pmatrix}_r \tag{11.47}$$

where suffix $r$ represents the reconstructed images used during NORICF synthesis.

## 11.10 Face recognition analysis using correlation classifiers

From Equation 11.40 and Equation 11.47 it is safe to comment that theoretically a delta-type correlation peak can be obtained due to correlation between $[\mathbf{H}_p^k]$ and $[\mathbf{H}_r^{jk}]$ when $j = k$. However, it may be noted that $[\mathbf{H}_p^k]$ and $[\mathbf{H}_r^{jk}]$ are block matrices and therefore these filters contain several classifiers depending on the values of $\alpha$ and $\beta$. Hence one-to-one correlation is needed to get the respective correlation planes as shown in Figure 11.9. From a set

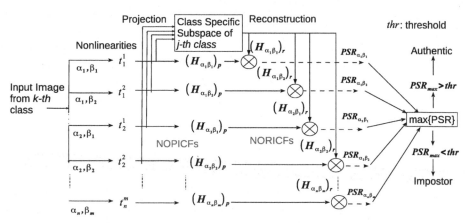

**FIGURE 11.9**: Block diagram of the system

of correlation planes, PSRs are evaluated and the maximum one is considered for decision. The detail of the filtering technique and decision making process regarding authentication is given in the block diagram shown in Figure 11.9. It is evident from Figure 11.9 that the multicorrelation approach is performed here for a single input image to evaluate the maximum PSR value. Although the illumination of the image mostly influences the magnitude spectrum, yet a major benefit is accrued by obtaining the phase spectrum.

As the poorly illuminated images contain more energy at low frequencies, the phase spectrum analysis of these images is a logical choice. As a delta-type correlation plane is desired for reducing classification errors, $\delta(m, n)$ is represented by a constant flat Fourier transform plane. This can be achieved if and only if phase-only NOPICF is identical to phase-only NORICF i.e. all the phases are cancelled out resulting in a constant flat spectrum. Hence, phase correlation between NOPICF and NORICF gives better results when compared to classical frequency domain correlation.

## 11.11 Test results

The PIE database contains two illumination subsets with 68 subjects and 21 images per subject. $640 \times 486$ pixel color images are converted into gray-scale images as the intensity is the main concern. All images are cropped to the size of $128 \times 128$. No other preprocessing is done. The YaleB (extended) database contains 38 different persons and for each person 64 differently illuminated gray-scale frontal face images of size $192 \times 168$ are present. These images are resized to $100 \times 100$. According to the lighting direction and camera position each individual's images are categorized into five subsets.

### 11.11.1 Comparative study on discriminating performances

In the first set of experiments, the discrimination ability of the filter between an authentic and impostor face image is tested. The phase-extended UMACE (PE-UMACE) and OTMACH[143] (PE-OTMACH) filters are designed with a typical set of training images and the multi-correlation approach is considered with one non-trained authentic image. In the multi-correlation approach, a set of correlation planes are developed corresponding to a test image for different values of $\alpha$ and $\beta$, while correlated with the designed PE-UMACE and/or PE-OTMACH. It is to be noted that when an image is multiplied with a scalar value $\alpha$, basically a linear operation is performed, i.e., the scaled image pixels will have the same dynamic range as the original one, if it is normalized within the range of gray level intensity [0−255] and no change in correlation plane will be observed. Hence throughout this study $\alpha$ is set to 1.

The change in correlation plane can be observed if image pixels are raised to $\beta$th amount as the image will be nonlinearly mapped with respect to the original one. In this study, values of $\beta$ set to 1 (to retain the original one) and $0.1, 0.2, 0.3$ (empirically) so that a narrow range of low intensity values are mapped into a wide range of high intensities and relatively high dynamic range of images and so also correlation filters can be achieved. Values such that $\beta > 1$ are ignored as low intensity images will be more darker and consequently discrimination capability of correlation filters will be lost.

A set of NOPICFs and NORICFs are evaluated corresponding to the test image for different values of $\alpha$ and $\beta$. Each NOPICF is correlated with a corresponding NORICF. From the set of response surfaces for all filters i.e. PE-UMACE, PE-OTMACH and the nonlinear one, the correlation plane associated with maximum PSR value is taken for making the decision of authentication. These correlation planes are shown in Figure 11.10. From Figure 11.10(a),Figure 11.10(c) and Figure 11.10(e) it is observed that the response surface corresponding to the nonlinear filtering method as shown in Figure 11.10(e) gives better discrimination ability compared to other filters.

The nature of the correlation plane corresponding to Figure 11.10(e) contains a sharp and distinct peak with high value and low sidelobes. This criterion is helpful in discriminating the authentic face images, which is reflected in Figs.11.10(b),(d),(f). The PSR values for impostors are shown with a surface boundary of PSR $= 10$. It may be noted that many of impostors are falsely accepted as authentic in the case of PE-UMACE and PE-OTMACH, as their PSR values are above 10. Considering Figure 11.10(b),(d), it is observed that fewer impostors are falsely accepted as authentic while the nonlinear technique is employed.

## 11.11.2   Comparative   performance   based   on   PSR distribution

To show the better verification performance, 20 sets of three randomly chosen training images are taken and the top-left corner image of Subset-5 is taken for testing. Obviously this test image is not included in the 20 sets during training of filters. In each training set the PSR value obtained from this method is greater than ten which is not so for other filters.

To test the verification performance of the present method the authentic PSR distribution is made with the help of sample order statistics. To develop this experiment 10 individuals are randomly chosen from 38 face images. For each individual 20 sets of training images are taken to synthesize all filters along with the present scheme. Hence for each individual, $20 \times 64$ authentic PSRs (APSRs) are obtained and then averaged. Having obtained APSR matrix of size $10 \times 64$ for 10 individuals, the normal distribution plot is made. This procedure is repeated for each standard filtering method along with the new one. Figure 11.11 shows the probability distribution plots for four different filtering methods.

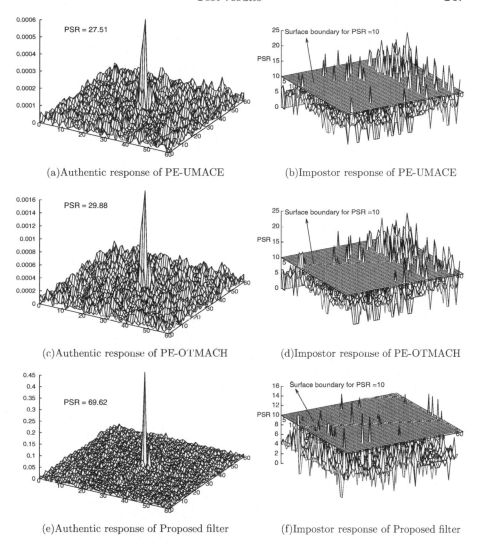

(a)Authentic response of PE-UMACE

(b)Impostor response of PE-UMACE

(c)Authentic response of PE-OTMACH

(d)Impostor response of PE-OTMACH

(e)Authentic response of Proposed filter

(f)Impostor response of Proposed filter

**FIGURE 11.10**: Response surfaces of (a),(c),(e) nontrained authentic and (b),(d),(f) impostor. Corresponding PSR values are given

For matrix APSR, the normal probability distribution plot displays a line for each column of APSR. The straight line indicates that the data originate from a normal distribution. Curvature indicates departure from normal distribution. Better linearity is obtained from 10 to 50 PSR values for the nonlinear filtering method. Therefore better normal distribution is achieved compared to other filters. At the higher end of the distribution, the PSR data for the present system are stretched out relative to the normal distribution. This indicates higher PSR values compared to other filters. Again

**TABLE 11.5**: The PSR value comparison of different filters corresponding to one unseen authentic image for 20 different training sets.

| Training Sets | MACH | UMACE | OTMACH | PEUMACE | Proposed |
|---|---|---|---|---|---|
| 9,8,45 | 6.7255 | 7.6867 | 8.2471 | 13.7647 | 21.1998 |
| 47,53,5 | 8.4226 | 10.1632 | 9.9003 | 17.3128 | 23.0935 |
| 35,57,19 | 4.4496 | 9.4087 | 8.1329 | 10.3363 | 18.7694 |
| 20,63,24 | 4.664 | 7.8933 | 6.7052 | 5.6747 | 13.0181 |
| 32,22,58 | 6.1131 | 11.6095 | 10.1757 | 10.0449 | 18.5307 |
| 3,2,34 | 5.0989 | 6.9466 | 5.325 | 11.183 | 24.2187 |
| 61,16,38 | 11.9977 | 13.9295 | 12.8193 | 22.0452 | 29.4638 |
| 44,39,8 | 6.24 | 6.0387 | 6.8613 | 10.5529 | 18.49 |
| 5,9,51 | 4.8421 | 7.3242 | 5.373 | 14.7871 | 21.1288 |
| 55,45,7 | 5.0781 | 6.7172 | 6.0718 | 13.1415 | 23.8946 |
| 21,8,43 | 0 | 5.9649 | 4.7964 | 8.5311 | 22.1234 |
| 30,8,63 | 0 | 5.3608 | 0 | 5.7914 | 21.1478 |
| 36,18,28 | 4.2209 | 9.79 | 8.8563 | 10.5253 | 19.3063 |
| 61,17,27 | 10.1446 | 13.5702 | 13.6904 | 15.6904 | 25.0827 |
| 60,5,17 | 7.1918 | 10.4287 | 9.9073 | 17.5891 | 22.3704 |
| 52,33,41 | 0 | 4.1415 | 0 | 7.6676 | 18.4477 |
| 39,8,34 | 6.499 | 0 | 0 | 5.2993 | f18.4176 |
| 59,18,42 | 6.0389 | 9.7192 | 8.0783 | 15.0277 | 20.6514 |
| 20,8,26 | 5.0756 | 7.0889 | 8.0338 | 5.5987 | 17.3505 |
| 19,6,26 | 5.7654 | 0 | 5.6837 | 0 | 20.9853 |

from the probability plot it is observed that the PSR values become zero for authentic images in the case of other correlation filters, which is not so for the present method.

## 11.11.3  Performance analysis using ROC

To further evaluate face verification performance of the nonlinear system both the databases are considered. Out of 21 face images from any one individual, only two images with image index 10 and 19 are taken for training. Another two training images are taken from YaleB subset-1. The logic behind training these images is that these images have no extreme variation of lighting. Hence the synthesized filters in both the training cases have no knowledge of extreme illumination variation of faces, as these are excluded from training. The performance of correlation filters are characterized, in terms of the $P_D$ and $P_{FA}$, with the help of ROC curves. To observe the robustness of the face recognition system the faces (excluding 10 and 19 in the case of PIE and excluding subset-1 in case of YaleB) are taken for testing and ROCs are plotted as shown in Figure 11.12(a),(b). The conventional binormal model is used to fit smooth ROC curves. From Figure 11.12(a),(b) it is

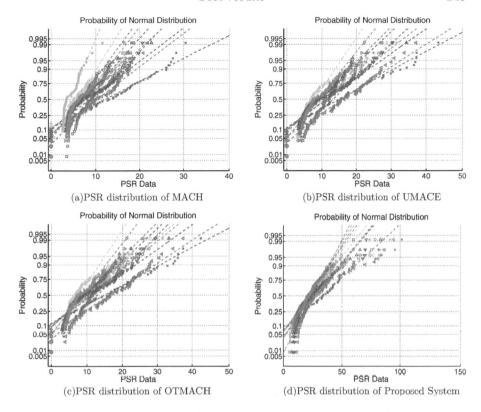

(a)PSR distribution of MACH

(b)PSR distribution of UMACE

(c)PSR distribution of OTMACH

(d)PSR distribution of Proposed System

**FIGURE 11.11**: Probability distribution of authentic PSRs for different filters

observed that the ROC corresponding to the nonlinear technique gives better traces of step function comparing to other filters.

Figure 11.12(c)-(f) shows different ROC plots for different set of training images (from YaleB) as (c) two random, (d) three random and from PIE (e) three random and (f) four random. Random images are taken 20 times and experimented over the whole database of YaleB and PIE. Having obtained authentic and impostor PSRs, $P_D$ and $P_{FA}$ are calculated and ROCs are plotted. From Figure 11.12(c)-(f) the ROC curves corresponding to the present method is approaching a step function indicating the better detection performance compared to the other filters.

Area under ROCs (AUC) are also calculated for Figure 11.12(c)-(f) so that the relative measurement of classification performance of different methods can be easily stated. As observed from Table 11.6, in each case highest AUC is obtained for a newly designed method. The 95% confidence interval is calculated from ROCs as given in Table 11.6. This indicates the interval in which the true AUC lies with 95% confidence.

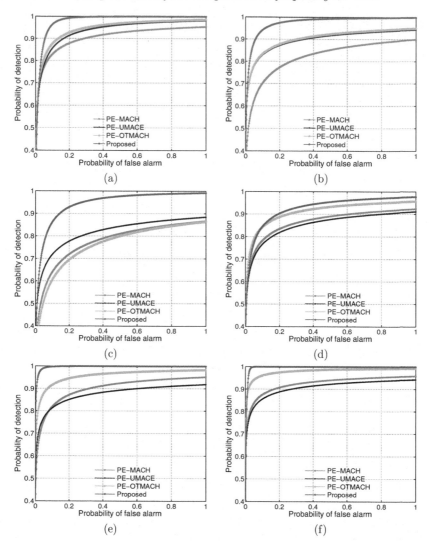

**FIGURE 11.12**: ROC plots for (a) training 10 and 19 from PIE; (b) training two images of subset-1 from YaleB. The improved recognition performance in terms of ROCs is shown with random images for (c), (d) YaleB and (e), (f) PIE. Comparisons are made with the phase-extended version of standard filters with a multicorrelational approach

## 11.11.4  Noise sensitivity

Noise sensitivity is further investigated as phase correlation is very much sensitive to noise. Under the inclusion of additive Gaussian noise the corrupted

**TABLE 11.6**: The area under the curve of ROCs corresponding to Figure 11.12(c)–(f) are given for comparison

| Training | Methods | | | |
|---|---|---|---|---|
| 2 random (CY) | PE-MACH | PE-UMACE | PE-OTMACH | Proposed |
| AUC<br>95% CI | 0.774<br>0.6919-0.857 | 0.82<br>0.74-0.892 | 0.76<br>0.674-0.844 | 0.943<br>f0.899-0.985 |
| 3 random (CY) | | | | |
| AUC<br>95% CI | 0.865<br>0.8-0.931 | 0.851<br>0.782-0.920 | 0.91<br>0.856-0.964 | 0.925<br>0.875-0.974 |
| 3 random (PIE) | | | | |
| AUC<br>95% CI | 0.898<br>0.841-0.955 | 0.874<br>0.811-0.938 | 0.955<br>0.917-0.99 | 0.996<br>0.985-1.0 |
| 4 random (PIE) | | | | |
| AUC<br>95% CI | 0.924<br>0.874-0.973 | 0.906<br>0.851-0.961 | 0.978<br>0.952-1.0 | 0.995<br>0.982-1.0 |

images are further tested. As noise can be characterized by variance, mean of Gaussian noise is set to 0 and different values of variance are considered as $0.001, 0.01, 0.1$. From AUC plots in Figure 11.13 it is observed that the nonlinear technique can tolerate illumination under noise when Figure 11.13(e) PIE faces are corrupted with noise variance up to 0.01 and Figure 11.13(f) YaleB faces are corrupted with noise variance up to 0.2, if $AUC = 0.9$ can be taken as sufficient recognition performance.

It has been seen from Figure 11.13(c),(d) the ROC curves degrade as the variance of noise is increased in both PIE and YaleB faces. This is due to the fact phase-only filters amplify the high frequency components and whenever noise is present it is also amplified and degrades the correlation planes. One solution can be made to tolerate illumination under additive noise by proper incorporation of a band-pass filter during phase only filter synthesis [209].

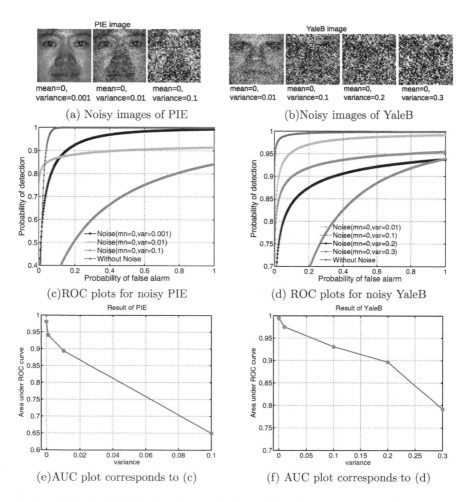

(a) Noisy images of PIE

(b)Noisy images of YaleB

(c)ROC plots for noisy PIE

(d) ROC plots for noisy YaleB

(e)AUC plot corresponds to (c)

(f) AUC plot corresponds to (d)

**FIGURE 11.13**: Sensitivity of nonlinear filtering technique with respect to additive Gaussian noise is shown for two different databases

# Chapter 12

# Landmark localization for face recognition

## 12.1   Introduction

Detection of facial landmarks is one of the first steps for face detection and recognition, emotion recognition, alignment of face images and many more [112]. The elastic bunch graph matching (EBGM) method provides a promising facial landmark detection where Gabor-jets are used to evaluate the features of landmarks and a face graph is developed to automatically select the test landmarks. In this chapter both EBGM and average synthetic exact filter applications towards landmark localization are given.

## 12.2   Elastic bunch graph matching

Elastic bunch graph matching (EBGM) is a feature-based face identification method. The algorithm assumes that the positions of certain fiducial points on the faces are known and stores the information about the faces by convolving the images around these fiducial points with 2D Gabor wavelets of varying size. The results of all convolutions form the Gabor jet for that fiducial point. EBGM treats all images as graphs (called face graphs), with each jet forming a node. The training images are all stacked in a structure called the face bunch graph (FBG), which is the model used for identification.

For each test image, the first step is to estimate the position of fiducial points on the face based on the known positions of fiducial points in the FBG. Then, jets are extracted from the estimated points and the resulting face graph is compared against all training images in the FBG, using Gabor jet similarity measures to decide on the identity of the person in the test image.

---

## 12.3 Gabor wavelets

Gabor wavelets are fundamental to the EBGM algorithm, which is a two dimensional form of Gabor wavelets. Wavelets are used, much like Fourier transforms, to analyse frequency space properties of an image, the difference being that the wavelets operate on a localized image patch, while the Fourier transform affects the whole image. Each wavelet consists of a planar sinusoid multiplied by a two dimensional Gaussian distribution. The sine wave is activated by the frequency information on the image, while the Gaussian ensures that the convolution result is dominated by the region close to the center of the wavelet. The wavelet specification follows an equation for its straightforward formulation and simplicity as

$$W(x, y, \theta, \gamma, \lambda, \sigma, \phi) = e^{-\frac{x'^2 + \gamma y'^2}{2\sigma^2}} \times \cos(2\pi \frac{x'}{\lambda} + \phi) \tag{12.1}$$

where

$$
\begin{aligned}
x' &= x\cos\theta + y\sin\theta \\
y' &= -x\sin\theta + y\cos\theta
\end{aligned}
\tag{12.2}
$$

Gabor wavelets can take a variety of different forms, usually having parameters that control the orientation, frequency, phase, size and aspect ratio. Hence, the wavelet specifications are governed by the following parameter selections:

1. $\theta$ specifies the orientation of the wavelet. This particular set uses eight different orientations over the interval 0 to $\pi$, i.e. $\theta \in \{0, \pi/8, 2\pi/8, 3\pi/8, 4\pi/8, 5\pi/8, 6\pi/8, 7\pi/8\}$.

2. $\lambda$ specifies the wavelength of the sine wave. This set starts at 4 pixels and continues to longer wavelengths at half-octave intervals, i.e. $\lambda \in \{4, 4\sqrt{2}, 8, 8\sqrt{2}, 16\}$.

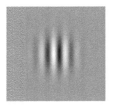  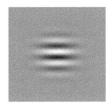 

**FIGURE 12.1**: Gabor wavelets for different orientations corresponding to $\theta$ of $0, \pi/4, \pi/2, 3\pi/4$

  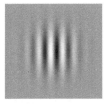 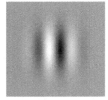

**FIGURE 12.2**: Gabor wavelets for different values of $\lambda$

3. $\phi$ specifies the phase of the sine wave. Typically Gabor wavelets are either even or odd. The even form of the sine wave corresponds to a cosine function; the odd form corresponds to a sine function. Even wavelets are thought to be the real part of the wavelet and the odd wavelets are thought to be the imaginary part of the wavelet. Therefore, a convolution with both phases produces a complex coefficient, i.e. $\phi \in \{0, \pi/2\}$.

4. $\sigma$ specifies the radius of the Gaussian. This parameter is usually proportional to the wavelength, such that wavelets of different size and frequency are scaled versions of each other, $\sigma = \lambda$.

5. $\gamma$ specifies the aspect ratio of the Gaussian. This parameter is included such that the wavelets could also approximate some biological models. The wavelets used are circular Gaussian, if $\gamma = 1$.

This parametrization yields eight orientations, five frequencies, and two phases for a total of eighty different wavelets. A coefficient is computed by convolving a location in the image with the wavelet kernel in Equation 12.1. Figure 12.3 shows forty wavelet filters with $\phi = \pi/2$ and Figure 12.4 illustrates another forty wavelet filters for $\phi = 0$. Figure 12.5 shows Gabor wavelets for different values of $\sigma$.

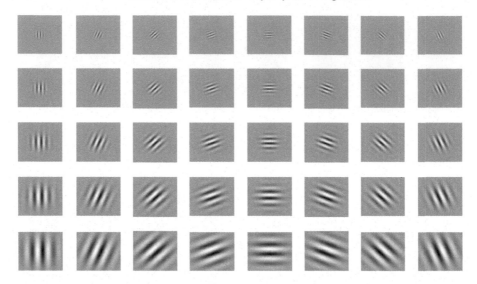

**FIGURE 12.3**: Forty wavelet filters with $\phi = \pi/2$. $\theta$ varies along the column and $\lambda$ varies along the row

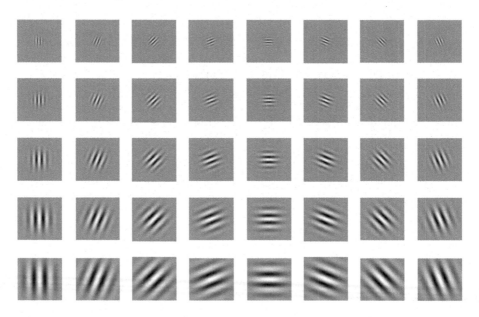

**FIGURE 12.4**: Forty wavelet filters with $\phi = 0$. $\theta$ varies along the column and $\lambda$ varies along the row

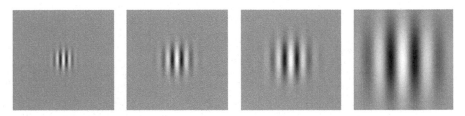

**FIGURE 12.5**: Gabor wavelets for different values of $\sigma$

## 12.4   Gabor jets

The Gabor wavelet transform yields a value for each wavelet at all locations of the image. Thus, with the standard parameters and discretized images it yields 80 (40 real + 40 imaginary) values at any pixel position. This set of values for a single pixel position is referred to as a jet $J$. Since a jet contains values from wavelets of different frequency and orientation, one can think of it as a local Fourier transform, and it is as such a representation of the local texture. It is in fact possible to reconstruct the image gray values from a jet in a small surrounding of its location, except for the mean value.

The Gabor wavelets come in pairs of cosine (real part) and sine (imaginary part) filters. Each filter by itself is relatively sensitive to a small shift, either of the image or of the pixel position in a stationary image. However, squaring and adding the responses of such pairs reduces the number of values to 40 and yields the local analog to a power spectrum, which still resolves frequencies and orientations but is insensitive to small shifts. In polar form, splits of all complex wavelet responses into amplitude and phase are given by

$$
\begin{aligned}
a &= \sqrt{a_{real}^2 + a_{imag}^2} \\
\psi &= \arctan\{\frac{a_{imag}}{a_{real}}\}, \qquad if, \qquad a_{real} > 0 \\
&= \pi + \arctan\{\frac{a_{imag}}{a_{real}}\}, \qquad if, \qquad a_{real} < 0 \\
&= \pi/2 \qquad if, \qquad a_{real} = 0, a_{imag} \geq 0 \\
&= -\pi/2 \qquad if, \qquad a_{real} = 0, a_{imag} < 0
\end{aligned}
\qquad (12.3)
$$

The array is stored as a $j$th complex wavelet pair. This internal representation of the feature for a given image is the Gabor jet for that feature. Use of only amplitudes is often advantageous, because it makes the jets more robust with respect to shifts and other transformations but at the price of not being able any more to reconstruct the local texture easily. Thus for localization and reconstruction one tends to use the full jets $J$ with phase information, while

**Wavelet masks**

Original Image                                    Magnitude

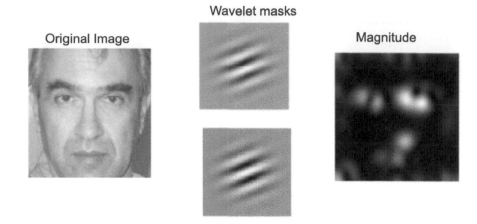

**FIGURE 12.6**: Convolution example with Gabor wavelets

for recognition purposes the reduced amplitude jets $|J|$ have proven to be more useful.

To acquire an accurate and comprehensive description of a feature in an image, it is necessary to convolve that location with a family of many different wavelets, typically having different frequencies and orientations. This multitude of convolution kernels leads to the notion of the Gabor jet. A two dimensional Gabor wavelet will respond to image features that are of the same orientation and size. The wavelet masks are centered over the correct location in the image and each corresponding value is computed by multiplying the pixel intensity with the mask value at that point and then summing up all individual contributions to the convolution. In order to compute both the real and the imaginary part of the wavelet, it is necessary to convolve the image with two masks that are out of phase by $\pi/2$, corresponding to the use of a sine and a cosine in the wavelet transform. Figure 12.6 shows the result of convolving an image containing a face with a real and imaginary wavelet. Figure 12.6 shows the original face image and the two wavelet masks that are used for convolution, and the rightmost image depicts the magnitude convolution values at each point. From the magnitude response it can be seen that the wavelets respond especially to the eyes and mouth corner position in the face and that magnitude values change rather slowly with the displacement from the center of the convolution.

## 12.5 The elastic bunch graph matching algorithm

The features are represented by Gabor jets, in this case referred to as model jets. The jets are extracted from images with manually selected landmark locations. The model jets are then collected in a data structure called a bunch graph. The bunch graph has a node for every landmark on the face. Every node is a collection model of jets for the corresponding landmark. The bunch graph serves as a database of landmark descriptions that can be used to locate landmarks in imagery.

The EBGM algorithm computes the similarity of two images. To accomplish this task, the algorithm first finds landmark locations on the images that correspond to facial features such as the eyes, nose and mouth. It then uses Gabor wavelet convolutions at these points to describe the features of the landmark. All of the wavelet convolution values at a single point are referred to as a Gabor jet and are used to represent a landmark. A face graph is used to represent each image. The face graph nodes are placed at the landmark locations, and each node contains a Gabor jet extracted from that location. The similarity of two images is a function of the corresponding face graphs.

Locating a face landmark has two steps. First, the location of the landmark is estimated based on the known locations of other landmarks in the image, and, second, the estimate is refined by extracting a Gabor jet from that image and comparing that jet to one of the models. Estimating the location of the other landmarks is easy based on the known location coordinates. For example, the eye coordinates are used to estimate the landmark location corresponding to the bridge of the nose. Because the bridge of the nose is relatively close to the eyes the estimate therefore can be very accurate. Location of the new landmark is then refined by comparing a Gabor jet extracted from the estimated point to a model jet. Now the new location along with the previous locations of the landmarks can be used to estimate the location of other landmarks and the process is iterated until all landmark locations are found.

The landmark location is refined by extracting a new jet from the estimated location of the landmark in the new image. The most similar jet is selected from the bunch graph and this jet then serves as a model. Both the new jet and model jet contain frequency information about the local image region around their extraction point. Using phase information stored in the two jets, it is possible to calculate a displacement of the new. The phase information of the novel jet is very similar to the phase information of the model jet.

Once the landmarks are located, a structure called a face graph is created where each node corresponds to a landmark. The landmarks are characterized by two parameters, a location in the image and a Gabor jet extracted from that location. After the face graph is created, the image is discarded, and the face graph becomes the internal representation of the face image. The face graph occupies less memory than the image, and computing the similarity of face graphs is much faster than computing the similarity of images. For this reason, an entire database of faces can be kept in memory, and new images can be identified rapidly.

The above procedure can be summarized in the following way. For the training step, the exact coordinates of these points are assumed to be known (usually hand-annotated by humans). Images are represented internally by the algorithm using spectral information of the regions around these features, which is obtained after convolving those portions of the image with a set of Gabor wavelets of varying size, orientation and phase. The results of the convolution for a specific position, the Gabor jets, are then collected for all fiducial points on a given image and aggregated together with the feature coordinates in that images face graph. Having applied this process to all images in the training set, all the resulting face graphs are concatenated in a stack-like structure called the FBG.

For the testing step, on the other hand, minimal information about the features is expected to be available. The algorithm constructs the test image's face graph by estimating the positions of fiducial points in an iterative manner, using the information stored in the FBG from previously estimated feature positions. After the face graph is constructed, it is compared against all members of the FBG to determine the closest match according to a given similarity metric. The identity of the person is thus established.

---

## 12.6   Application to face recognition

Once we have the means to generate and compare image graphs, recognition of faces in identical poses is relatively straightforward. Matters become more complicated when trying to recognize faces across different poses. For face recognition the following steps are followed.

1. Building a face graph: The first step to bootstrap the system is to define the graph structure for the given pose. Thus, we take the first image and manually define node locations on the face that are easy to localize, such as the corners of the eyes or mouth, the center of the eyes, the tip of the noise, some points on the outline, etc. Thus the first face graph is generated.

2. Building a face bunch graph: The single face graph defined above can be viewed as a bunch graph with just one instance in it. It can be matched onto the second face image, but if the first two face images are not very similar, the match may be of poor quality. The bunch graph with two instances is then matched onto the third image, to have a third instance for the bunch graph. By repeating this process, the bunch graph grows, and as it grows the match onto new images gets more and more reliable.

3. Building the model gallery of graphs: Since we now have a bunch graph that provides sufficient quality for finding the node locations in a new face, we can process the remaining images fully automatically. We are now in the position to perform face recognition on a new probe image.

4. Building the probe graph: First we need to create a graph for the probe image. This process works exactly as done for the model images.

5. Comparison with all model graphs: The image graph is compared with all model graphs, resulting in similarity values. These form the basis of the recognition decision. Notice that this does not require EBGM anymore; only the graphs are compared according to the similarity function.

6. Recognition: The model graph with the highest similarity with the image graph is the candidate to be recognized. However, if the best similarity value is relatively low, the system has to decide on several ambiguities decided by the robustness of the system.

---

## 12.7 Facial landmark detection

Accurate registration of face images is an important first step in face recognition, and one common way of establishing face registration is by finding eyes, where an eye is one of the important facial landmarks. A simple and robust correlation filter called average exact synthetic filter (ASEF)[154] is discussed for eye localization.

### 12.7.1 ASEF correlation filter

ASEF filters differ from other correlation filters in that the convolution theorem is exploited to greatly simplify the mapping between the input training image and the output correlation plane. In the Fourier domain the correlation operation becomes a simple elementwise multiplication, and therefore each corresponding set of Fourier coefficients can be processed independently. The resulting computations also naturally account for translational shifts in the spatial domain. As a result the entire correlation output can be specified for each training image. ASEF filters are trained using response images that specify a desired response at every location in each training image. This response typically is a bright peak centered on the target object of interest. One consequence of completely specifying the correlation output is a perfect balance between constraints and degrees of freedom for each training image, and therefore an exact filter is determined for every

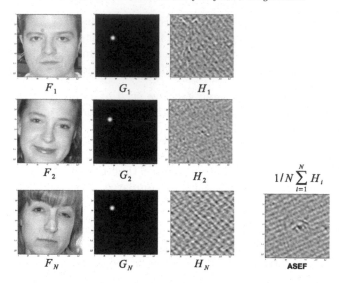

**FIGURE 12.7**: Construction of ASEF. The final correlation filter ASEF is produced by taking the average of many exact filters.

training image. ASEF filters also provide much more freedom when selecting training images and when specifying the synthetic output. A benefit is that the training images need not be centered on the target. For each training image, we specify the desired filter output and may place the peak wherever the target appears. Because the correlation peak moves in locksteps with the targets in the training images, all the exact filters are consequently registered by inverting the correlation process. This increases training flexibility, allowing to customize the desired response for each training image. For example, training images may have multiple targets per training image as long as the synthetic output contains multiple corresponding peaks.

### 12.7.2 Formulation of ASEF

Figure 12.7 shows the pictorial representation of the process of constructing an ASEF. The training pairs $F_i$; $G_i$ consist of a training image and associated desired correlation output. The correlation image $G_i$ is synthetically generated with a bright peak at the center of the target, in this case the left eye, and small values everywhere else. Specifically, $G_i$ is defined as a two dimensional Gaussian centered at the target location $(x_i, y_i)$ with radius $\sigma$

$$G_i(x, y) = \exp -\frac{(x - x_i)^2 + (y - y_i)^2}{\sigma^2} \tag{12.4}$$

The correlation plane in response to the $i$th image can be expressed in frequency domain as,

$$\mathbf{G}_i = \mathbf{F}_i \mathbf{H}_i^* \tag{12.5}$$

where, $\mathbf{F}_i$ and $\mathbf{H}_i$ are Fourier transformed $F_i$ and $H_i$. Equation 12.5 leads to the solution of the exact filter as,

$$\mathbf{H}_i^* = \frac{\mathbf{G}_i}{\mathbf{F}_i} \tag{12.6}$$

where the division is an elementwise division between the transformed target output $\mathbf{G}_i$ and the transformed training image $\mathbf{F}_i$.

It can be seen from Figure 12.7 that the exact filters $H_1$, $H_2$ and $H_N$ do not appear to have a structure that would respond well to an eye but instead are specific to each training image. To produce a filter that generalizes across the entire training set, the average of the multiple exact filters is computed. Averaging emphasizes features common across training examples while suppressing unreliable features of single training instances. This is visually evident in the final ASEF shown in the bottom row of Figure 12.7.

In particular, the exact filter can be thought of as a weak classifier that performs perfectly on a single training image. A summation of a set of weak classifiers outperform all the component classifiers and, more importantly, if the weak classifiers are unbiased, their summation converges upon a classifier with zero variance error. The idea leads to the formulation of ASEF, which can be expressed as,

$$\mathbf{H}_{\mathrm{ASEF}} = \mathbf{H}^* = \frac{1}{N} \sum_{i=1}^{N} \mathbf{H}_i^* \tag{12.7}$$

In spatial domain the ASEF filter is obtained as,

$$\mathrm{H}_{\mathrm{ASEF}} = \mathrm{FFT}^{-1} \mathbf{H}^* \tag{12.8}$$

The real part of $\mathrm{H}_{\mathrm{ASEF}}$ is shown in the bottom row of Figure 12.7. It is suggested in [154] that ASEF filters perform best when trained on as many images as possible.

## 12.8 Eye detection

The ASEF filter is tested on the Caltech face database. Faces were initially found in all images using the OpenCV face detector (from www.opencv.com). Detected faces are cropped and resized to $128 \times 128$. With these cropped images, eyes are detected by, OpenCV eye detector and at left eye's position the Gaussian filters are generated. No further intensity normalization or preprocessing has been done; however, better results can be obtained if

**FIGURE 12.8**: Eye detection using ASEF under different challenging conditions.

preprocessed as suggested in [154]. ASEF was trained on the full $128 \times 128$ image tile. Figure 12.8 shows some left eye detection of test images using an ASEF filter.

## 12.9 Multicorrelation approach

### 12.9.1 Design of landmark filter(LF)

From Equation 10.2 it is observed that the correlation plane, in response to a test image, in frequency domain is obtained by elementwise multiplication of the Fourier transformed test image and the synthesized filter. In the other way round it may be considered as if the correlation plane and the test image are known beforehand, and the filter $\mathbf{F}$ can be generated. In case of designing a landmark filter, in the training phase the position of the left eye, right eye, nose tip and mid mouth is known. It is expected that after correlation operation between test image and landmark filter/indexLandmark filter, the generated correlation plane $\mathbf{G}$ must show a distinct peak at the position of the landmark due to the shift-invariant property.

Hence the design aspect of the landmark filter consists of a training set $\{x_i, g_i\}$, where $x_i$ is the training image and $g_i$ is the desired correlation output plane. This $g_i$ is generated synthetically by assuming a 2D Gaussian curve cenetred at position $(r, c)$ of one of the landmarks, expressed mathematically

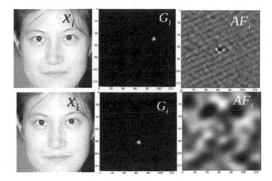

**FIGURE 12.9**: Top row shows the generation of AF for the right eye and bottom row shows the generation of AF for nose tip.

as

$$g(i,j) = \frac{1}{\sqrt{2\pi}\sigma} exp \left\{ \frac{(i-r)^2 + (j-c)^2}{\sigma^2} \right\} \tag{12.9}$$

The filter $\mathbf{F}_i$ is developed corresponding to $\mathbf{X}_i$ and $\mathbf{G}_i$ as

$$\mathbf{F}_i^* = \frac{\mathbf{X}_i}{\mathbf{G}_i} \tag{12.10}$$

Filter $\mathbf{F}_i$ is developed with a single training image. To make a robust representation of the $i$th landmark, $N$ number of training images are required. To produce a filter that generalizes across the entire training set, the average of $N$ such landmark filters is computed as

$$\mathbf{AF}_i = \frac{1}{N} \sum_{i=1}^{N} \mathbf{F}_i \tag{12.11}$$

Averaging emphasizes features common across training examples while suppressing idiosyncratic features of single training instances [154]. Production of $\mathbf{F}_i$ is pictorially given in Figure 12.9. Figure 12.9 illustrates the synthesis of two landmark filters.

In order to finally derive the closed form solution of landmark filter (**LF**), some modification is done. A metric can be introduced in order to force the correlation outputs from all images in the training set to match the average of the correlation outputs from some exemplars. Instead of using $\mathbf{af}_i$,[1] $(\mathbf{af}_i - \beta\mathbf{m})$ is introduced to modify the **AF** filter solution so that the relative influence of average image is incorporated in the filter solution. Here $\beta$ is the controlling parameter depending on what the relative influence of the mean image is exploited.

The exemplar $(\mathbf{af}_i - \beta\mathbf{m})$ is now the $i$th training pattern with part of

---

[1]**af** is the lexicographic version (of size $d \times 1$) of **AF** of size $d_1 \times d_2$ and $d = d_1 \times d_2$.

the mean subtracted. Hence it is desirable for all patterns in the training set to follow this exemplar's behaviour. This can be done by forcing every pattern in the training set to have a similar correlation output plane to an ideal correlation output shape **g**. To find the **g** that best matches all these exemplar's correlation output planes its deviation from their correlation plane is minimized. This deviation can be quantified by the average squared error (ASE) as,

$$
\begin{aligned}
\text{ASE} &= \frac{1}{N} \sum_{i=1}^{N} |\mathbf{g}_i - \mathbf{g}|^2 \\
&= \frac{1}{N} \sum_{i=1}^{N} (\mathbf{g}_i - \mathbf{g})^+ (\mathbf{g}_i - \mathbf{g})
\end{aligned}
\tag{12.12}
$$

where

$$
\mathbf{g}_i = (\mathbf{A\bar{F}}_i - \beta\mathbf{\bar{M}})^* \mathbf{lf}
\tag{12.13}
$$

and $\mathbf{A\bar{F}}_i = diag\{\mathbf{af}_i\}$, $\mathbf{\bar{M}} = diag\{\mathbf{m}\}$.

$\mathbf{lf}^2$ is the desired filter vector corresponding to 2D landmark filter **LF** To find the optimum shape vector $\mathbf{g}^{opt}$ the gradient of ASE in Equation 12.12 is set to zero and $\mathbf{g}^{opt}$ is obtained as

$$
\nabla_g(\text{ASE}) = \frac{2}{N} \sum_{i=1}^{N} (\mathbf{g}_i - \mathbf{g}) = 0
\tag{12.14}
$$

or

$$
\mathbf{g}^{opt} = \frac{1}{N} \sum_{i=1}^{N} \mathbf{g}_i
\tag{12.15}
$$

Hence the optimal shape vector is formulated as,

$$
\begin{aligned}
\mathbf{g}^{opt} &= \frac{1}{N} \sum_{i=1}^{N} (\mathbf{A\bar{F}}_i - \beta\mathbf{\bar{M}})^* \mathbf{lf} \\
&= \left\{ \frac{1}{N} \sum_{i=1}^{N} \mathbf{A\bar{F}}_i - \beta\mathbf{\bar{M}} \right\}^* \mathbf{lf} \\
&= \left\{ (1 - \beta)\mathbf{\bar{M}} \right\}^* \mathbf{lf}
\end{aligned}
\tag{12.16}
$$

Now the average similarity measure can be modified as the measure of dissimilarity of the training images to $(1-\beta)\mathbf{\bar{M}}^*\mathbf{lf}$ and can be mathematically

---

[2]**lf** is the lexicographic version of **LF** of size $d_1 \times d_2$.

expressed as [195]

$$ASM_{new} = \frac{1}{N} \sum_{i=1}^{N} |\mathbf{A\bar{F}}_i^* \mathbf{h} - (1-\beta)\mathbf{\bar{M}}^* \mathbf{lf}|^2$$

$$= \frac{1}{N} \sum_{i=1}^{N} (\mathbf{A\bar{F}}_i^* \mathbf{h} - (1-\beta)\mathbf{\bar{M}}^* \mathbf{lf})^+ (\mathbf{A\bar{F}}_i^* \mathbf{lf} - (1-\beta)\mathbf{\bar{M}}^* \mathbf{lf})$$

$$= \frac{1}{N} \sum_{i=1}^{N} (\mathbf{lf}^+ \mathbf{A\bar{F}}_i - \mathbf{lf}^+ (1-\beta)\mathbf{\bar{M}})(\mathbf{A\bar{F}}_i - (1-\beta)\mathbf{\bar{M}})^* \mathbf{lf}$$

$$= \frac{1}{N} \sum_{i=1}^{N} \mathbf{lf}^+ (\mathbf{A\bar{F}}_i - (1-\beta)\mathbf{\bar{M}})(\mathbf{A\bar{F}}_i - (1-\beta)\mathbf{\bar{M}})^* \mathbf{lf}$$

$$= \mathbf{lf}^+ \{\frac{1}{N} \sum_{i=1}^{N} (\mathbf{A\bar{F}}_i - (1-\beta)\mathbf{\bar{M}})(\mathbf{A\bar{F}}_i - (1-\beta)\mathbf{\bar{M}})^*\} \mathbf{lf}$$

$$= \mathbf{lf}^+ \mathbf{\bar{S}}_{new} \mathbf{lf} \qquad (12.17)$$

where,

$$\mathbf{\bar{S}} = \frac{1}{N} \sum_{i=1}^{N} (\mathbf{A\bar{F}}_i - (1-\beta)\mathbf{\bar{M}})(\mathbf{AF}_i - (1-\beta)\mathbf{\bar{M}})^* \qquad (12.18)$$

In addition to the above performance criteria, the desired $\mathbf{lf}_k$ must yield a large peak at the correlation plane at the position of the $k$th landmark. This criteria is met by maximizing the average correlation height (ACH) criterion as follows:

$$\mathrm{ACH} = \frac{1}{N} \sum_{i=1}^{N} \mathbf{af}_i^+ \mathbf{lf} = \mathbf{m}^+ \mathbf{lf} \qquad (12.19)$$

Hence to make ACH large while minimizing $ASM_{new}$, the filter is designed to maximize

$$J(\mathbf{lf}) = \frac{|ACH|^2}{ASM_{new}} = \frac{|\mathbf{m}^+ \mathbf{lf}|^2}{\mathbf{lf}^+ \mathbf{\bar{S}}_{new} \mathbf{lf}} \qquad (12.20)$$

The closed form solution of the above equation can be obtained as [140]

$$\mathbf{lf} = \mathbf{S}^{-1} \mathbf{m} \qquad (12.21)$$

Equation 12.21 represents the final form of the landmark filter. In this study four such landmark filters are designed to locate four landmark points in a face image.

## 12.9.2 Landmark localization with localization filter

Having obtained the $\mathbf{LF}_k, k = \{1, 2, 3, 4\}$ the $k$th landmark can be simply found out by cross correlation of $\mathbf{LF}_k$ and the extracted face part from the detected skin region. Extracted faces are resized to a fixed dimension of 128 × 128. The output correlation plane provides the maximum value or distinct peak at the position of the $k$th landmark.

Figure 12.10 illustrates the correlation planes with bright spots corresponding to the landmark position. Figure 12.11 illustrates a 3D correlation plane where the distinct peak is visible corresponding to the midmouth position in response to a test image. The positions of the maximum value at the correlation planes are recorded and mapped into the original image using coordinate information.

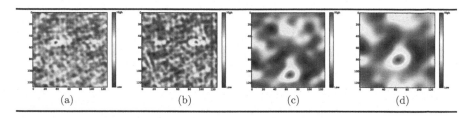

|     |     |     |     |
| --- | --- | --- | --- |
| (a) | (b) | (c) | (d) |

**FIGURE 12.10**: Correlation planes obtained for (a) left eye, (b) right eye, (c) midmouth, (d) nose tip.

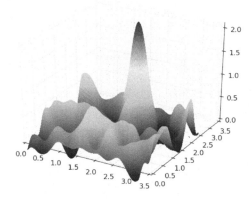

**FIGURE 12.11**: 3D mesh plot showing a peak at mid-mouth position

## 12.10   Test results

The above algorithm is evaluated on the CalTech database. Some pose variation results are also shown for the LFW(Labeled Faces in the Wild) database. Gray-scale images contain only the intensity information and hence are discarded for evaluation purposes. Performance of the proposed scheme is categorized in different modules depending on the test images. The CalTech face database contains 450 images which is further divided into six different categories: 1) high light, 2) poor light, 3) complex, 4)pose variation, 5) scale variation and 6) normal. Figure 12.12 illustrates the face detection and landmark localization results on a single face image under high lighting conditions, face with glass, with complex backgrounds and scale variations.

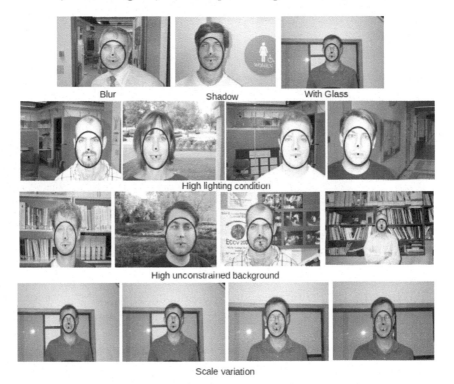

**FIGURE 12.12**: Performance of the landmark filter under different conditions

Figure 12.13 also illustrates the system performance under different pose variations with very low resolution images taken from the LFW database. Performance of this strategy is evaluated on multiple faces as shown in Figure 12.13. It has been observed that in the case of multiple face cases ,false positives (FPs) are generated. A detected face is a correct detection if the detected locations of the eyes, the mouth and the ellipse bounding a human face are found with a small amount of tolerance; otherwise it is called an FP. Ellipse fitting is based on the position of landmark locations. The major axis of an ellipse is the approximate Euclidean norm between the left eye and mid mouth position, where the minor axis is evaluated by determining the norm between the left and right eye with some tolerance values. The center of

**FIGURE 12.13**: Performance of the landmark filter under pose variation and low resolution and bottom row illustrates localization for multiple faces

**TABLE 12.1**: Performance evaluation on CalTech. FP: number of false positive; NOI: number of total images

| Category | Face detection | Landmark localization |
|---|---|---|
| | FP(NOI) | FP (NOI) |
| High frontal light | 2(26) | 8(26) |
| Poor light | 16(21) | 16(21) |
| Complex background | 6(18) | 7(18) |
| Pose variation | 4(22) | 8(22) |
| Scale variation | 6(17) | 7(17) |
| Normal | 2(346) | 3(346) |
| % detection rate | 92% | 89.12% |

the ellipse is guided by the nose tip position. Performance evaluation of the proposed method is summarized in Table 12.1 where both correct detection rate (%CR) and false positive rate (%FP) is given. The detection rate is computed by the ratio of the number of correct detections in a gallery to that of all human faces in the gallery. Landmarks of all images in the database are collected manually. Having automatically detected the landmark position, the Euclidean distance is calculated with respect to the stored coordinates. If the distance exceeds a predefined value, the detected landmark is discarded. Figure 12.14 illustrates the error generated between actual and obtained landmark coordinates for 50 randomly taken images. A hard threshold of error value is set to 10, above which the detected landmarks are discarded.

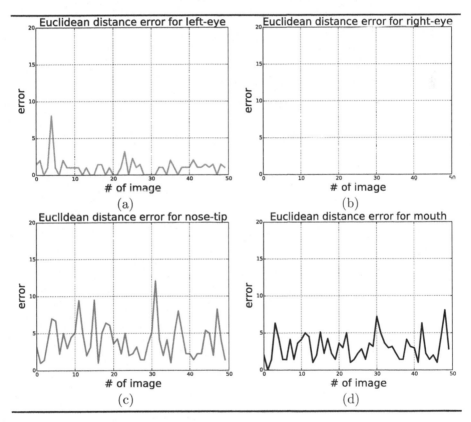

**FIGURE 12.14**: Error plot for different landmark localization with respect to their actual positions.

According to Table 12.1 it is observed that almost 90% detection accuracy is achieved. Poor detection rate is observed in the case of low lighting conditions. This is due to the fact that the design method does not have any information about illumination variation.

# Chapter 13

# Two-dimensional synthetic face generation using set estimation

## 13.1   Introduction

Synthetic generation of face images for the purpose of face recognition has been explored in recent times. Two dimensional (2D) to three-dimensional (3D) reconstruction and generation of new face images of various shapes and appearances had received attention. The popular solution to the problem is proposed by 2D and 3D modeling of faces. 3D models include face mesh frames, morphable models and depth map-based models, where one needs to incorporate high quality graphics and complex animation algorithms. Flynn et al. [85] provided a survey of approaches and challenges in 3D, multi-modal 3D and 2D face recognition. 3D head poses are derived from 2D to 3D feature correspondences [86]. Face recognition based on fitting a 3D morphable model is also proposed with statistical texture [87]. Some significant works are discussed in Table 2.5. Four main approaches for 2D modeling are active appearance models (AAMs) [80], manifolds [210], geometry-driven face synthesis methods [68] including face animation[82] and expression mapping techniques [108],[84].

2D to 2D face reconstruction was initially developed by Cootes et al. [80] and their active appearance, generated from 2D face images, was one of the powerful methods. Multi-view face reconstruction in 2D space is done by manifold analysis. Geometry-driven face synthesis [108] and expression mapping techniques are also useful in 2D face generation. Table 2.4 provides comparative appraisal on advantages and disadvantages of some useful techniques of face synthesis. New face images are generated either from a model or from some functional properties.

## 13.2    Generating face points from intraclass face images

Face class for a particular person $P$ can be represented as a set consisting of infinitely many face images. Face images of $P$ differ from each other mainly because of different expressions resulting from the muscle movements in different portions of the face. Movement of eyebrows, twitching of nose, muscle movement in cheeks, movement of lips, opening and closing of mouth alone and different combinations thereof are some examples of variations in facial expressions. For a person $P$, if his/her face is continuously photographed and stored, then for every two images in the set, there exists a path in the set that joins those two images. That is, there exists a path containing infinitely many images joining those two images.

Let $x_1, x_2, ...., x_n$ be the points in the training set for the $i$ th class and let $\xi_i$ be the estimated radius for this class. Thus, the estimated set can be written as, $\bigcup_{j=1}^{n}\{x : d(x, x_j) \leq xi_i\}$. For the generation of two or more points, one can use the information on suitable edges, which can be put in the form of an algorithm, as stated below:

**Algorithm:**

**Step1**: Find minimum spanning tree (MST) for each face class $i$.

**Step 2**: For MST of a face class and for each edge joining two points $x$ and $y$, find intermediary points $p$, given by

$$p = \lambda x + (1 - \lambda)y, \lambda \in (0, 1) \tag{13.1}$$

**Step 3**: Reconstruct the face corresponding to every new face point generated.

The number of intermediate points and the corresponding values for $\lambda$ are decided on the basis of the requirements. The process of selection of new face points by intra class feature generation is illustrated in Figure 13.1. In the figure, two classes each having five face points in the training set, denoted by $\circ$, are used. The intermediate face points denoted by $\star$ are generated having the features of both the nodes of an edge.

### 13.2.1    Face generation using algorithm with intraclass features and related peak signal to noise ratio

For each class, a minimal spanning tree (MST) is constructed with the face points of the probe set. For generating new face images, the considered value

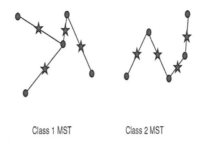

Class 1 MST          Class 2 MST

**FIGURE 13.1**: Intraclass feature generation

for $\lambda = 0.5$ (for AR dataset) and 4 new images for each class are generated using the method stated in Algorithm 1. While applying Algorithm 1 on the training set, MST for each class has 4 edges. Thus, 4 new images are generated for the same class. The total number of images that could be generated is therefore 160. Some representative generated face images are shown in column 3 of Figure 13.2. These generated face images are sharing the properties of the images in columns 1 and 2 of Figure 13.2, where 7 representative images from 160 images are shown.

Peak signal to noise ratio (PSNR) is used to judge the quality of an image with respect to a given image. A PSNR value greater than 30 dB, in general, indicates the closeness of the two considered images. Quality of the generated image is taken as acceptable, if the PSNR is greater than 20dB [211]. PSNR value of each one of 160 new images, taken 4 images per class from 40 classes, is found to be greater than 35 dB. Some remarks on result set 1 can be made as:

1. In the generated images of rows 1 to 5, the eyes are visible from spectacles. Thus, the effect of sunglasses is not present in the generated images.

2. Illumination direction is compensated in the generated faces. In rows 1, 3 and 5, the first image is left illuminated whereas the second one is right illuminated. The resultant images have no pronounced left or right illumination directions.

3. In rows 6 and 7, the generated face images inherited sunglasses (one feature) from one face and a muffler (other feature) from the other face. Generated faces possess both the artifacts used by the two persons from which the face is generated.

4. Note that the number of new images created for two different classes is the same. This is because of (i) the number of training sample points from each class is 5, (ii) the MST has 4 edges for each class and (iii) the image is generated for each edge of MST.

**FIGURE 13.2**: Result set 1. For each row, the image in the third column is generated as a synthetic image from the images of columns 1 and 2.

The algorithm is also applied on the training set of the FIA dataset. The purpose of this experiment is to show the smooth transition from one expression to another. For each edge in MST, 7 images are generated by taking the values of $\lambda = 0.2, 0.3, 0.4, 0.5, 0.6, 0.7$ and $0.8$. Seven such representative results are shown in Figure 13.3.

In each row of images in Figure 13.3, the first and the last images are the input images and the rest are generated images. The generated faces correspond to the intermediate face points of the edges joining the two end-training points of the MST. Changes in facial expressions are indicated as (a)gradual transformation of eyelids and lip, (b) Gradual changes in nose and mouth, (c)eyes and mouth are slowly opened, (d) opened mouth and teeth are closing with corresponding movement of eyelids, (e) expression are changed to normal, (f) smiling faces are returning to normal and (g) expression of surprise is changing to pleasant face. These sets of images can also be used to produce a video, showing the changes in facial expressions.

---

## 13.3   Generating face points from interclass face images

New face images for a face class which possesses features of face images from other classes can be generated. For example, the person in one face class may never wear spectacles whereas in the other face class, the corresponding person wears spectacles. Another example may be cited, where the eyes of a person in one class may be open, whereas the eyes of a person in another face class may be closed.

Note that the maximal edge weight of the MST of $n$ points is $\xi_i$. Every edge weight in the MST is less than or equal to $\xi_i$ . Thus, for every edge in MST joining two points, say $x$ and $y$ in multi dimensional space, the corresponding discs with centers $x$ and $y$ and radii $\xi_i$ intersect. The intersection of these discs indicates sharing of the properties of the two vectors $x$ and $y$ in the intersection region. On the other hand, if there is no edge joining $x$ and $y$, then the discs do not intersect, and thus the properties do not share. Since the objective is to generate new images with possibly additional information, the region corresponding to the intersection of two discs is important. In an MST of $n$ points i.e., $(n\text{-}1)$ edges, every point on each edge belongs to the intersection region of two discs. The middle point of an edge is likely to possess the properties of both the images equally.

In the Figure 13.4, two face classes are formed in the face space with two intersecting discs along with their radii $\xi_1$ and $\xi_2$. New face images are generated from the thick line portion (-) of the intersection region A and B. The line joining the probe discs are centered at $z_1$ and $z_2$. The generated face points must incorporate features from two different classes. In reality, these features can also be the artifacts used in faces.

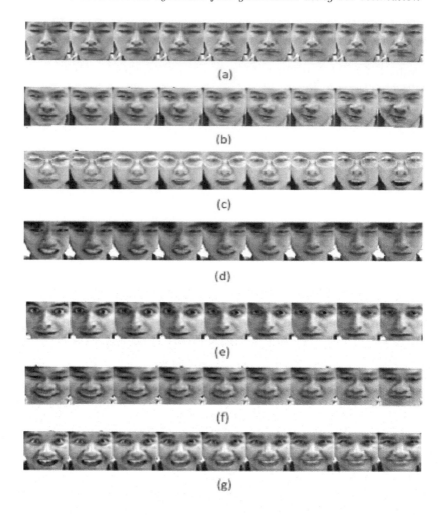

**FIGURE 13.3**: Result set 2. For each row, the smooth changes of images from one end point of the edge of the MST to another are shown.

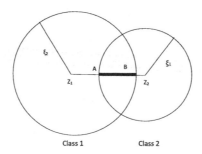

**FIGURE 13.4**: Interclass feature generation

The problem of generating new images may be tackled in two ways. One way is to identify features in a target face image which are desired to be seen on the new face image of a person. The next task is to find another face image having the said features. A combination of the features of both face images at the same location can then be made. It is also necessary to judge whether the combination is meaningful or not. The second way is to take combinations of face images and decide whether the changes incorporated in the generated face image are meaningful. The term meaningful is used in the restricted sense. The result may be an image with very high degree of noise or may not have usual features of a face and thus those are not meaningful. In both the approaches, combinations of faces need be considered and judgements about the combinations are to be made.

The first approach proposed is heuristic in nature. As explained later in this chapter, for some cases the generated images may be distorted. Therefore, an alternative approach by generating face points taking the features from other face classes in the face space is needed. The approach is stated in the algorithm and the steps are as follows:

**Step 1**: Determining the value of radius $\xi$ of a face class

For each class, MST of the respective $N$ vectors is calculated and its maximal edge weight is found. Let the maximal edge weight of the MST of the $i$-th class is denoted by $\xi_i$. If the reduced set for the $i$-th class after dimensionality reduction is denoted by $z_{1i}, z_{2i}, \ldots, z_{Ni}$, then the estimated set for the class $i$ be denoted by $B_i$, and is obtained as,

$$B_n = \bigcup_{j=1}^{N} \{ y \in \Re^m : d(y, z_{ji}) \leq \xi_i \}, i = 1, 2, M. \tag{13.2}$$

**Step 2**: For every class $i$, find its nearest face class $j$ using the following steps.

Let $a_{ik} = Min\{d(z_{ji}, z_{wk}), 1 \leq l, w \leq N\}, 1 \leq i, k \leq M, i \neq k$. For every $i$, $1 \leq i \leq M, \exists j, 1 \leq j \leq M, j \neq i$, $a_{ij} = min(a_{ik}) : k \neq i$. Such $a_{ij}$ are denoted as $b_{ij}$. Without loss of generality, $z_1 \in \{z_{1i}, z_{2i}, \ldots z_{Ni}\}$ and $z_2 \in \{z_{1j}, z_{2j}, \ldots z_{Nj}\}$ are such that $b_{i,j} = d(z_1, z_2)$.

**Step 4**: If $b_{ij} \geq \xi_i + \xi_j$, then go to the next value of $i$. Otherwise, join the two points $z_1$ and $z_2$ by a line segment and generate a face point from the intersecting region of the discs of radii $\xi_i$ and $\xi_j$ centered at $z_1$ and $z_2$, respectively, and falling on this line segment (see Figure 13.4)

**Step 4.1**: The geometrical formulation used to generate points on the line segment joining the points A and B is given by

$$A = \frac{(b_{ij} - \xi_j)z_2 + \xi_j z_1}{b_{ij}}; B = \frac{(b_{ij} - \xi_i)z_1 + \xi_i z_2}{b_{ij}} \tag{13.3}$$

where, $d(z_1, z_2) = b_{ij}, d(z_2, A) = \xi_j, d(z_1, B) = \xi_i, d(z_1, A) = b_{ij} - \xi_j$ and $d$ is the Euclidian distance.

**Step 5**: Reconstruct the new faces corresponding to the generated face points.

The above-mentioned linear combinations are used to generate the face points on the line segment.

## 13.3.1 Face generation with interclass features

In the experimental part of face generation with inter class features, the generated images have the dominant properties of the training set. Some additional face properties like artifacts or expressions are also inherited in the generated faces. Two sets of images are shown in Figure 13.5 and those in Figure 13.6 are from the AR dataset. Both sets of images are generated using Algorithm 2, where the features of two intersecting classes are mixed. The newly generated face images are the convex combinations of the intersecting classes. All generated faces are from the intersecting region and on the line joining the points A and B as shown in Figure 13.4.

In image set 2, the smooth changes of two different faces of one person to another can be clearly seen. The resulting set is generated from the AR dataset with a training set of neutral faces without any artifacts. The corresponding indices for the faces are 1, 2, 3, 4 and 5. Number of faces in test set consist of all images from other classes. This way of dataset formation is used to show the efficiency of the algorithm. Similar results on FIA data set for expressions can be obtained.

Algorithm 2 is applied on face images of two different classes as shown in Figure 13.5, where eleven representative results are considered. In each row, the first image is of the person whose characteristics are to be transferred. The second image is of the person receiving these characteristics. The third image is the image of the second person after receiving the said characteristics. Thus, the third image is the generated face from the first two images. The generated image corresponds to the middle point of the line segment falling in the intersection region and it joins the centers of the two discs.

PSNR values for the generated images are also computed with respect to the original images to show the quantitative outcomes on the quality of the generated images. All PSNR values range from 25 dB to 35 dB. From the PSNR values, the second image is found to be closer to the generated image than the first. The number of such intersections between the classes is not more than 15. PSNR values establish the justification of generating a face image from two different intersecting classes.

Some remarks, however, can be made on the results:

1. As found from the images, one can view that the spectacles are added in each of the morphed faces of rows 1 to 8. The sunglasses as the dominant feature are also added to the face images.

2. In rows 9 to 11, the face images of column 1 are *mufflered* face images,

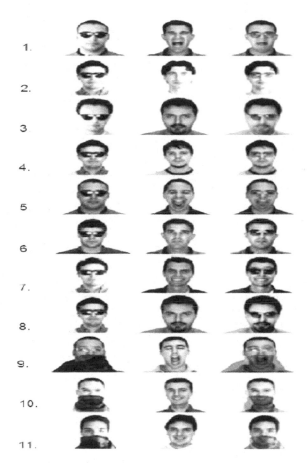

**FIGURE 13.5**: Result set 3. For each row, the image in the third column is generated as a synthetic image from the images of columns 1 and 2.

where each generated face image (column 3) has a muffler. Note that the hidden portion of the face can be viewed in the generated face.

3. It is not possible to generate new images for every pair of face classes. Non intersecting classes do not generate meaningful new faces. Thus the number of new images added to two different face classes in the data base may not be the same.

4. It is apparent from the images of column 3, that each row of the generated face is the face of the person shown in column 2 but not the one shown in column 1. To find mathematically, the class identity of the generated faces, the Euclidean distance has been calculated and a neural net classifier is used to classify each generated face. Since the main intention is to add artifacts to the images of column 2, the generated subset of the class of the image in column 2 is naturally expected. In 100% of cases the images are found to be classified to the proper class i.e., the class of image 2 in each row.

One may also be interested in seeing the intermediate images when the face of one person is changed to the face of another person. The result (set 4), shown in Figure 13.6, depicts such a transition for five representative cases. Similar results may be obtained for the other cases too.

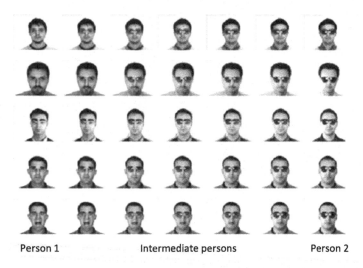

Person 1              Intermediate persons              Person 2

**FIGURE 13.6**: Image set 4. In each row, the smooth change of faces of intersecting classes is shown

### 13.3.2 Rejection of the non-meaningful face and corresponding PSNR test

Any linear combination of the two face images may not provide meaningful face images. An algorithm has been developed with the notion that the intersecting classes only can share the facial features. Some rejected faces which were generated from the non intersecting classes are shown in Figure 13.7. From the subjective judgement, it can be stated that these images are highly distorted and classifying any image to one of the existing classes is a difficult task and may not be possible. In each such case, a PSNR value is found to be less than 10 dB. This again justifies the contention that new face images should not be generated from non intersecting face classes.

**FIGURE 13.7**: Result set 5. For each row, the image in the third column is the generated distorted face from the first two images, where the corresponding face classes are non-intersecting.

## 13.4 Generalization capability of set estimation method

In the previous sections, new images are generated from two training images, where those two images may be either from the same class or from different classes. In this section, attempts are made to generate test images from the training images. Consequently, several intermediate images between a training image and a test image are generated. Theoretically, using the method of set estimation one can generate *uncountably many* images under a specific radius.

If $x$ is a vector in $n$ dimensional Euclidean space, and $y$ is a vector whose Euclidean distance from $x$ is $r$, then $y$ can be exactly generated from $x$ by suitably using $(n-1)$ angle variables $\theta_1, \theta_2, \theta_3, ....., \theta_{n-1}$, where each $\theta_i \in (0, 2\pi]$. In fact, if $\theta_1, \theta_2, \theta_3, ....., \theta_{n-1}$ are chosen appropriately, then

$$y = x + r \begin{bmatrix} \sin\theta_{n-1}\sin\theta_{n-2}\cdots\cdots\sin\theta_1 \\ \cos\theta_{n-1}\sin\theta_{n-2}\sin\theta_3\cdots\sin\theta_1 \\ \cos\theta_{n-2}\sin\theta_{n-3}\sin\theta_4\cdots\sin\theta_1 \\ \cos\theta_{n-3}\sin\theta_{n-4}\sin\theta_5\cdots\sin\theta_1 \\ \cos\theta_2\sin\theta_1 \\ \cos\theta_1 \end{bmatrix} \tag{13.4}$$

Additionally, many images can be generated corresponding to points on the line segment joining $x$ and $y$. These images reflect the smooth transition from $x$ to $y$. After the reduction in the number of features, if the nearest neighbor of a test image $y$ is $x$, then $y$ is classified to the class of $x$. Let $\xi$ denote the value of the threshold, calculated using set estimation procedure for the class of $x$, and the distance between $x$ and $y$ be $\gamma$. Then, one of the following cases may arise.

1. The classes of $x$ and $y$ are the same and $\gamma \leq \xi$.

2. The classes of $x$ and $y$ are the same and $\gamma > \xi$.

3. The classes of $x$ and $y$ are different and $\gamma \leq \xi$.

4. The classes of $x$ and $y$ are different and $\gamma > \xi$.

Case 1 occurs for at least 90% of test images. Since $\gamma \leq \xi$, $y$ falls in the disc of radius $\xi$ and thus $y$ can be exactly generated from $x$. Six more equi-spaced synthetic images are generated by varying the distance from $x$ from $0$ to $\gamma$ without changing the values of $\theta_i$. These images are present neither in the training set nor in the test set of the data. PSNR values with respect to $y$ are increased to 35 dB as the distance from $x$ is increased to $\gamma$. Figure 13.8 shows the training image $x$, test image $y$ and the intermediate images. The images are generated using Equation 13.4.

Case 2 occurs when $\gamma > \xi$, and y does not fall in the closed disc of radius $\xi$ centered at $x$. Thus, one can generate an approximation of $y$ lying on the boundary of the disc and the distance between the approximated image and the actual image is the least among all images lying in the disc. Since $\gamma > \xi$, the generated images are not expected to be close to $y$ compared to the images generated in the previous case. In Figure 13.9, some examples of $x$, $y$ and the closest approximation for $y$ are provided. PSNR values with respect to $y$ are found to be around 15 dB for the synthetic images.

Case 3 does not usually occur, since $y$ and $x$ are from different classes as well as $\gamma \leq \xi$. Additionally, the occurrence of this case signifies that there are

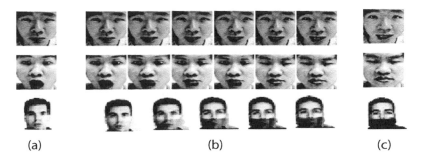

(a)                                  (b)                                  (c)

**FIGURE 13.8**: Result set 6. Training image, test image and intermediary images, where training and test images are from the same class where (a) is the face images of person 1, (b) are the intermediate faces generated and (c) faces of person 2.

some common characteristics between $x$ and $y$. In this case, $y$ can be exactly generated from $x$. Generating images with the considered datasets, this case did not occur even once.

**FIGURE 13.9**: Result set 7. Training image, test image and intermediary images, where training and test images are from the same class

Case 4 condition occurs when a point is forcibly classified to a class and the information from the training set is insufficient for classification. In this case $y$ is misclassified and $y$ is less similar to points from its own class than from another class. This represents a case of a training set being a less representative set of the overall variations of its class. Here, nothing can be said about which image is closest to $y$. Generating images with the considered datasets, this case has not occurred. For the datasets considered, this case has not occurred.

Considering all cases, it may be seen that the proposed set estimation method can generate test images perfectly in at least 90% of the cases and also has generated approximations in around 9% of the cases.

## 13.5    Test of significance

Test of significance is carried out to compare the recognition rate for different training sets consisting of the original face image along with the newly generated face image. *Z-test* for the equality of proportions is employed to check the performance of the proposed method. The test statistic is given by

$$a = \frac{p_1 - p_2}{\sqrt{p(1-p)((\frac{1}{n_1}) + (\frac{1}{n_2}))}} \qquad (13.5)$$

where $p_1 = \frac{x_1}{n_1}$ and $x_1$ is the number of correctly classified images out of $n_1$ images using the PCA and NN classifier and where $p_2 = \frac{x_2}{n_2}$, $p = \frac{x_1+x_2}{n_1+n_2}$. The observed values of $a$ are greater than 3 for all cases and the values lie outside not only of the 95% confidence interval $(-1.96, 1.96)$ but also outside the 99% confidence interval $(-2.575, 2.575)$. Thus, the hypothesis that the two proportions are same is rejected.

An experiment is conducted to check whether the synthetic images alone can perform classification better than the original face images. For this purpose, different training sets are made with a different number of synthesized images. The experiment is carried out on a different number of images as training sets. The results on two such sizes are shown in Table 13.1. In the first such training set, for each edge in MST, exactly one synthesized image is generated (i.e., the corresponding $\lambda$ value =0.5), resulting in a total 4 number of training images for each class. In the second training set, for each edge in MST, four synthesized images are generated where the corresponding $\lambda$ values are 0.2, 0.4, 0.6 and 0.8. Thus the number of training images for each class is sixteen. This experiment is carried out on all dimensionality reduction methods. The results on PCA and 2DPCA are shown in Table 13.1. In general, it has been observed that the recognition rates are improving as the size of the training set is increased for every dimensionality reduction method.

This chapter deals with the techniques on face generation and face verification with the estimated face classes. The features are shared by using inter- and intraface classes. The selection of meaningful faces from the generated faces is successfully done using estimated disc radius. It is shown that the distorted faces are automatically rejected for both FIA and AR datasets. In the statistical approach mentioned, many new face images are generated in a face class with the help of training sample points from the same class, resulting in face images with different expressions as well as different artifacts. The convex combinations considered between those points form the edge of MST. The smooth variation in face images from one class to another class is also shown. The representation of a face class would be more accurate if more points are present in the training set corresponding to that class. PSNR value is used to judge the quality of a generated image with respect to a given

**TABLE 13.1**: Variation of recognition rates, where the training set consists of only the generated images

| Face database | Method applied | λ Values | Number of images generated from each face class using the proposed method | Recognition rate with the generated faces |
|---|---|---|---|---|
| FIA | PCA | 0.5 | 41 | 90 |
| | | 0.2, 0.4, 0.6, 0.8 | 16 | 96.5 |
| AR | 2DPCA | 0.5 | 4 | 90.5 |
| | | 0.2, 0.4, 0.6, 0.8 | 16 | 96 |

image and the values for all generated images are found to lie in the interval of 25 dB to 40 dB. This result reflects good and acceptable performance of the proposed methods.

The other aspect of the method is to classify the newly generated faces into their respective classes, where nearest neighbour classifier is used. As an extension, the whole process is applied and tested with dimensionality reduction techniques like KPCA, 2DPCA, PCA-LDA, and the classification results are acceptable. When the generated images alone are used in the training set, satisfactory recognition rates are also obtained. Reasonably low number of training images are used; however, the results are expected to improve if a training set of large size is used. The generalization capability of generating the test images using the set estimation method is shown for images which are not present in the training dataset.

---

MATLAB code for synthetic face generation

---

```
\texttt{% finding recognition rate between test and
   %training (existing plus synthetic) images
of FERET face database
   clc
   clear all

   % STEP 1: Read training and test set

   % INPUT
   nPointRange=1; % number of points taken
within one mst edge should be varied in this range
   nClass=50; % total number of classes
   nImg=11; % total number of images in each class
   m=64;n=43; % size of the image is mXn
after being scaled by 1/3
   totImgId=[1 2 3 4 5 6 7 8 9 10 11];
   nImgTrain=6; % number of images in each train class
   nImgTest=5; % number of images in each test class
   nCombination=20;

   load("feret.mat");
% load the dataset in the variable "set"

   % find all combinations of training and test set
   trainImgId=nchoosek(totImgId,nImgTrain);
   for row=1:size(trainImgId,1)
       testImgId(row,:)=setdiff(totImgId,trainImgId(row,:));
   end
   resultSynth=[];resultClub=[];

   %nCombination=size(trainImgId,1);
% maximum number of combinations
   for row=1:nCombination
       tic
       col=1;
       fprintf("\n *******
%d th COMBINATION of train and test set *******\n",row);
   trainImgId(row,:)
   testImgId(row,:)
   % Extract train and test set from "set"
   train=extractSet(set,1:1:nClass,trainImgId(row,:),nImg);
   test=extractSet(set,1:1:nClass,testImgId(row,:),nImg);

       % STEP 2: Perform PCA on train
and test set and find out the recognition rate
```

```
using nearest-neighbor.
      [projTrain projTest]=pca(train,test,size(train,2));
      fprintf
      ("\n\n Recognition rate without using synthesized images :\n")
      [correct,wrong,rate]=nearNeighbor(projTrain,projTest,nClass)
      resultSynth(row,col)=rate;
resultClub(row,col)=correct;resultClub(row,col+1)
=wrong;resultClub(row,col+2)=rate;


      % STEP 3: For each class of the training set
 form an MST using Prim's algorithm, considering each
      % image of the class as a node.
      % Now, generate "nPoint" number of
EQUIDISTANT synthetic image within each edge.
      % Combine the synthetic images with
 existing ones to form the clubbed set and
      % find out the recognition rate between
 clubbed set and test set using nearest-neighbour.
      for nPoint=nPointRange
          col=col+1;
          projSynth=[];
% projection matrix of synthetic set
        projClub=[];
  % projection matrix of clubbed (existing + synthetic) set
          for class=1:nClass
              imaginary=[];
% synthetic space projection for single class "class"
              img=0;
% not necessary : used as image index
only for  writing purpose
              lowImg=(class-1)*nImgTrain+1;
% lower bound of image in class "class"
              upImg=lowImg+nImgTrain-1;
 % upper bound of image in class "class"
% already existing space projection for single class "class"
              exist=projTrain(:,lowImg:upImg);
              existDist=edistMat(exist);
              [cost,next]=prim(existDist,1);
% apply prims algo to get the MST
              for node=2:nImgTrain
% next(1) is always zero
                  nextNode=next(node);
 % node and nextNode are two vertices of the mst edge
                  lambda=0.0;
                  for point=1:nPoint
                      lambda=lambda+1/(nPoint+1);
  % this is a synthetic node (image point)
 between node and nextNode
restore1=lambda*exist(:,node)+(1-lambda)*exist(:,nextNode);
```

```
                imaginary=[imaginary restore1];
imwrite(restore4,path);
                end % next point
                end % next node
                threshold(class)=max(cost)/2.0;
                projSynth=[projSynth imaginary];
                projClub=[projClub exist imaginary];
            end % next class
            fprintf("\n WHEN NUMBER OF POINTS (IMAGES)
 TAKEN WITHIN ONE EDGE OF MST IS = %d",nPoint)
fprintf("\n\n the recognition rate using original plus
synthesized images :\n")
    [correct,wrong,rate]=nearNeighbor(projClub,projTest,nClass)
resultClub(row,col+2)=correct;resultClub(row,col+3)=wrong;
resultClub(row,col+4)=rate;
        end % now vary number of synthetic points
        fprintf(" time taken in
%d th combination out of total %d combinations is",row,nCombination);
        toc
    end % next combination

  resultClub

y=resultClub(:,2)
z=resultClub(:,5)

ybar=mean(y);zbar=mean(z);
stdDev_y=sqrt(var(y))
stdDev_z=sqrt(var(z))
n1=length(y)
n2=length(z)
est=sqrt(var(y)/n1+var(z)/n2)
g1=var(y)/n1;
g2=var(z)/n2;
df_nume=(g1+g2)^2*(n1-1)
df_deno=g1^2+g2^2
df=df_nume/df_deno
t=(ybar-zbar)/est}
```

# Chapter 14

# Datasets of face images for face recognition systems

## 14.1    Face datasets

In order to compare different techniques of face detection and recognition and also to assess how well those methods work, several face images databases have been developed. The number of databases and the size of the gallery and probe sets used for testing a system are indicative of the robustness of the methods. Several face databases are thus generated by different research groups which provide as many variations as possible on their images. Each database is designed to address specific challenges covering a wide range of scenarios. While existing publicly available face databases contain face images with a wide variety of poses, illumination angles, gestures, face occlusions and illuminant colors, these images have not been adequately annotated, thus limiting their usefulness for evaluating the relative performance of face detection algorithms. For example, many of the images in existing databases

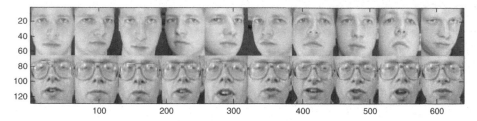

**FIGURE 14.1**: Sample images from ORL dataset

are not annotated with the exact pose angles at which they were taken. Descriptions of some of the datasets are given in the following subsections. The entire set of images, as well as the annotations and the experimental results, is being placed in the public domain and made available for download over the worldwide web.

In order to compare the performance of various face recognition algorithms presented in the literature there is a need for a comprehensive, systematically annotated database populated with face images that have been captured (1) at a variety of pose angles (to permit testing of pose invariance), (2) with a wide variety of illumination angles (to permit testing of illumination invariance) and (3) under a variety of commonly encountered illumination color temperatures (permit testing of illumination color invariance). PIE and AR face datasets represent one of the most popular 2D face image database collection. FERET represents a good testing framework if one needs large gallery and probe sets, while CMU is more indicative when pose and illumination variations are considered. In contrast, the AR face dataset is the only one which provides occluded face images.

### 14.1.1   ORL dataset

The ORL dataset contains face images of 40 persons, with 10 images of each. For most subjects, the images are shot at different times and with different lighting conditions. However, the images are taken against a dark background. Difficulties with the dataset are (1) limited number of people, (2) illumination conditions are not consistent from image to image and (3) the images are not annotated for different facial expressions, head rotation or lighting conditions.

### 14.1.2   OULU physics dataset

The OULU physics dataset [212] includes frontal color images of 125 different faces. Each face was photographed 16 times, using one of four different light sources (horizon, incandescent, fluorescent, and daylight) in combination with one of four different camera calibrations (color balance

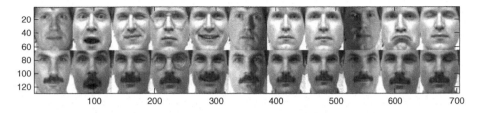

**FIGURE 14.2**: Sample images from Yale dataset

settings). The images are captured under dark room conditions against a gray screen. The spectral reflectance (over the range from 400 nm to 700 nm) is also measured at the forehead, left cheek and right cheek of each person with a spectrophotometer. The spectral sensitivities of the R, G and B channels of the camera, and the spectral power of the four sources, are also recorded over the same spectral range.The drawbacks of the database are (1) although the images are captured under a good variety of illuminant colors, there are no variations in the lighting angle and (2) all face images are basically frontal, with some variations in pose angle and distance from the camera.

### 14.1.3   XM2VTS dataset

The XM2VTS dataset [213] is basically a video face database. It consists of 1000 GBytes of video sequences and speech recordings taken from 295 subjects at 1 month intervals over a period of 4 months (4 recording sessions). Significant variability in appearance of clients, such as changes of hairstyle, facial hair, shape and presence or absence of glasses is present in the recordings. During each of the four sessions a speech video sequence and a head rotation video sequence are captured. This database is designed to test systems designed for multimodal (video with audio) testing of facial and voice features. It does not include any information about the image acquisition parameters, such as illumination angle, illumination color or pose angle.

### 14.1.4   Yale dataset

The Yale dataset [66] contains frontal gray-scale face images of 15 people, with 11 face images of each subject, giving a total of 165 images. Lighting variations include left light, center light, and right light. Images of subject have variations including with glasses and without glasses. Facial expression variations include normal, happy, sad, sleepy, surprised and winking. Its limitations are (1) limited number of people, (2) while the face images in this database are taken with three different lighting angles (left, center and right) the precise positions of the light sources are not specified, (3) since all images are frontal, there are no pose angle variations and (4) ambient lighting conditions are also not specified.

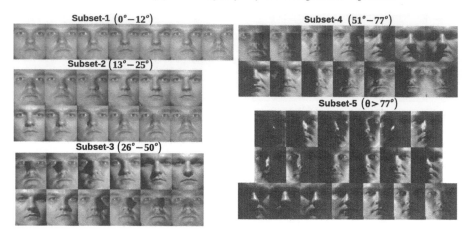

**FIGURE 14.3**: Sample images from Yale B dataset

### 14.1.5   Yale-B dataset

The Yale-B dataset [207] contains gray-scale images of 10 subjects with 64 different lighting angles and nine different poses angles, for a total of 5760 images. The images are captured with an overhead lighting structure which is fitted with computer-controlled xenon strobe lights. For each pose, images were captured of each subject at a rate of 30 frames/sec, over a period of about 2 seconds. Pose 0 is a frontal view, in which the subject directs his/her gaze directly into the camera lens. In poses 1 to 8 the subject is gazing at specified angles. However, the dataset has (1) limited number of subjects. (2) the background in these images is not homogeneous and is cluttered and (3) the different pose angles in these images are not precisely specified.

### 14.1.6   MIT dataset

The MIT dataset contains 16 subjects. Each subject is photographed 27 times, while varying head orientation. The lighting direction and the camera zoom are also varied during the sequence. The resulting 480 x 512 gray-scale images are then filtered and sub sampled by factors of 2, to produce six levels of a binary Gaussian pyramid. The six pyramid levels are annotated by an $X \times Y$ pixel count, which ranged from 480×512 down to 15×16. Some of the difficulties with this database are (1) scale variations, lighting variations and pose variations are not very extensive and are not precisely measured and (2) apparently the subjects are moving between pictures.

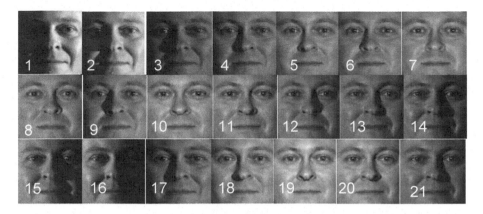

**FIGURE 14.4**: Sample images from PIE dataset

### 14.1.7 PIE dataset

The CMU-PIE (pose illumination and expression) dataset [208] contains images of 68 subjects with 13 different poses, 43 different illumination conditions and 4 different facial expressions, for a total of 41,368 color images of size 640 × 486. Two sets of images are captured one set with ambient lighting and the other set without ambient lighting. Unfortunately, (1) clutters are visible in the backgrounds of these images and (2) exact pose angle for each image is not specified.

### 14.1.8 UMIST dataset

The UMIST dataset [214] consists of 564 gray-scale images of size 220×220, of 20 people of both sexes from various races. Various pose angles of each person are provided, ranging from profile to frontal views. However, (1) absolute pose angle is not provided for each image and (2) information about the illumination used, either its direction or its color temperature, is not provided.

### 14.1.9 PURDU AR dataset

The PURDU AR dataset [128] contains over 4000 color frontal view face images of 70 men and 56 women that are taken during two different sessions separated by 14 days. No restrictions on clothing, eyeglasses, make-up or hair style are imposed upon the participants. Controlled variations include facial expressions (neutral, smile, anger and screaming), illumination (left light on, right light on, all side lights on), and partial facial occlusions (sunglasses or a scarf). However, the placement of light sources or their color temperature

**FIGURE 14.5**: Sample images from AR dataset

is not specified. Moreover, the placement of the two light sources produces objectionable glare in the spectacles of some subjects.

## 14.1.10    FERET dataset

The FERET dataset [129] is used frequently and contains face images of over 1000 people. It was created during the period from 1993 through 1997. The database is assembled to support government-monitored testing and evaluation of face recognition algorithms using standardized tests and procedures. The final set of images consists of 14051 gray-scale images of human heads with views that include frontal views, left and right profile views and quarter left and right views. It contains many images of the same people taken with time-gaps of one year or more, so as to record changes in facial features. This is important for evaluating the robustness of face recognition algorithms over time. However, (1) it does not provide a very wide variety of pose variations and (2) there is no information about the lighting used to capture the images.

## 14.1.11    Performance    evaluation    of    face    recognition    algorithms

One could divide the evaluations of performance into two categories or types: technological and operational. Each of these evaluation types focuses on different aspects and uses different approaches. Ideally, the evaluation of a system that serves a particular purpose starts with a technology evaluation, followed by an operational evaluation. The purpose of a technology evaluation is to determine the underlying technical capabilities of a particular system against a database of face images collected under previously determined conditions. Technology in this context is understood to be the different types of facial recognition algorithms and related hardware. The evaluation is normally performed under laboratory conditions using a standardized dataset that was compiled in controlled conditions ensuring control over pose, illumination, background and resolution. The results of evaluations are used by developers to refine their systems under the test conditions. However, the evaluation can also be used by potential customers to select the most appropriate technology for a particular application. In this context sometimes, operational evaluations are also performed. Operational evaluations aim to study the impact of specific systems on the organization of work flow and the achievement of operational objectives. It is also imperative to conduct scenario evaluations to evaluate the overall capabilities of the entire system in a real-world environment and population. This would include the involvement of an image-capturing component.

The lack of publicly available data on evaluations of face recognition and detection systems is a major concern for users. Sometimes, it might be necessary to consider the most prominent example of evaluations done in vendor tests (FRVT) and the facial recognition grand challenge (FRGC) conducted by the National Institute of Standards and Technology (NIST). This is often helpful in interpreting the performance of a system for a particular application scenario.

## 14.2 FERET and XM2VTS protocols

Before 1996, there did not exist a common face recognition technology (FRT) evaluation protocol that included large databases and standard evaluation methods. The U.S. army formed the FERET program in 1993, and the database was collected over three years (1993 to 1996). It consists of 14,126 images of 1199 individuals with the main objective of measuring the performance of FRT in a framework that models a real-world setting.

The XM2VTS project was formed in 1999 by the European advanced communications technologies and services project and deals with access control by multimodal identification of human faces. The objective of XM2VTS was to improve recognition performance by combining the modalities of face and voice. Its database contained five shots of each of 37 objects. The database is divided into three parts: a training set, an evaluation set and a test set. The protocol intended to evaluate the performance of vision- and speech-based person authentication systems.

Though now obsolete, the FERET and XM2VTS evaluation technologies had a tremendous impact on progress in face recognition research and ultimately led to FRGC and FRVT.

## 14.3    Face recognition grand challenge (FRGC)

The primary goal of the face recognition grand challenge (FRGC) was to promote and advance face recognition technology and help in developing new face recognition techniques and prototype systems while increasing performance by an order of magnitude. The FRGC was structured around challenge problems that were designed to challenge researchers to meet the FRGC performance goal. The FRGC was open to face recognition researchers and developers in companies, academia and research institutions, by proposing progressively difficult challenge problems. Each challenge problem consisted of a data set of facial images and a defined set of experiments. One of the impediments to developing improved face recognition is the lack of data. The FRGC provides data consisting of 50,000 recordings of high resolution still images, 3D images and multi-images of a person. The traditional method for measuring the size of a face is the number of pixels between the centers of the eyes. In current images there are 40 to 60 pixels between the centers of the eyes (10,000 to 20,000 pixels on the face). In the FRGC, high resolution images consist of facial images with 250 pixels between the centers of the eyes on average. Usually, preprocessing a facial image is conducted to correct for lighting and pose prior to being processed through a face recognition system. The preprocessing portion of the FRGC also measures the impact of preprocessing algorithms on recognition performance.

Out of 50,000 recordings of the dataset, the training partition is designed for training algorithms and the validation partition of 4,003 subject sessions is for assessing performance of an approach in a laboratory setting. A subject session is the set of all images of a person taken each time a person's biometric data are collected and consists of four controlled still images, two uncontrolled still images and one three-dimensional image. The controlled images are full frontal facial images taken under two lighting conditions and with two facial expressions (smiling and neutral). The uncontrolled images are taken in varying illumination conditions in open spaces. Each set of uncontrolled images contains two expressions, smiling and neutral. The 3D images are taken under controlled illumination conditions and consist of both a range and a texture image.

The FRGC data distribution consists of three parts. The first is the FRGC data set. The second part is the FRGC biometric experimentation environment (BEE) with an XML-based framework for describing and documenting computational experiments. The BEE distribution includes all the data sets for performing and scoring the six experiments. The BEE will allow the description and distribution of experiments, recording of the raw results of an experiment, analysis and presentation of the raw results and documentation of the experiment, all in a common format. This is the first time that a computational-experimental environment has supported a challenge problem in face recognition or biometrics. The third part is a set of baseline algorithms for experiments 1 through 4. With all three components, it is possible to run experiments 1 through 4, from processing the raw images to producing receiver operating characteristics (ROCs).

FRGC distribution consists of six experiments. In experiment 1, the gallery consists of a single controlled still image of a person and each probe consists of a single controlled still image. Experiment 2 studies the effect of using multiple still images of a person on performance. Experiment 3 measures the performance of 3D face recognition with the gallery and the probe set consists of 3D images of a person. Experiment 4 measures recognition performance from uncontrolled images, where the gallery consists of a single controlled still image, and the probe set consists of a single uncontrolled still image. Experiments 5 and 6 compare the performance with 3D and 2D images. In both experiments, the gallery consists of 3D images. In experiment 5, the probe set consists of a single controlled still and in experiment 6, the probe set consists of a single uncontrolled still.

The most significant conclusions one might draw from the results of the FRGC are (1) uncontrolled environments are still a significant problem, (2) 3D recognition using both shape and texture does not necessarily provide better results than high quality 2D images and (3) though the performances are improving, they are still lower than expected.

---

## 14.4 Face recognition vendor test (FRVT)

The face recognition vendor test (FRVT) is a series of large scale independent evaluations for face recognition systems realized by the National Institute of Standards and Technology in 2000 [215], 2002 and 2006 [216]. Previous evaluations in the series were the Face Recognition Technology (FERET) evaluations conducted in 1994, 1995 and 1996. The primary objective of the FRVT was to provide performance measures for assessing the ability of automatic face recognition systems to meet real-world requirements and has an impact on future directions of research in developing automated face recognition systems. FRVT 2002 was the high computational intensity test (HCInt) which consisted of 121,589 operational images of 37,437 people. From these data, real-world performance figures on a very large data set were computed on performance statistics for verification and identification. Some important conclusions from FRVT 2002 are worth mentioning, such as (1) face recognition systems do not appear to be sensitive to normal indoor lighting changes, (2) recognition from video sequences is not better than from still images, (3) three-dimensional morphable models substantially improve the ability to recognize non-frontal faces and (4) males are easier to recognize than females and younger people are harder to recognize than older people.

The widely reported FRVT of 2002 was followed by the FRVT 2006 evaluation [217]. Independent assessments were performed by NIST and was sponsored by by multiple U.S. Government agencies such as the Department of Homeland Security, the Director of National Intelligence, the Federal Bureau of Investigation, the Technical Support Working Group and the National Institute of Justice. A standard dataset and test methodology was employed so that all participants were evenly evaluated. The test environment was called the Biometric Experimentation Environment (BEE). For evaluation tasks, high-resolution 2D still images (5 to 6 megapixels) and 3D images (both a shape and texture) were used. Simultaneous evaluation of iris recognition technology was also conducted. For evaluation of algorithms, performance was compared to human performance. Pre-processing algorithms that compensate for pose and illumination were also evaluated. It may be pointed out that the vendors of FRT often use results from the technology evaluations (FRVT, FRGC, etc.) to make claims about their products more generally without providing the context of such evaluations. This leads to misleading conclusions about the efficacy of the technology [218].

---

## 14.5    Multiple biometric grand challenge

It seems that face recognition alone may not provide total security in high risk zones. So attempts are made to integrate other facial parameters, such as iris for identification of a person. To effect this philosophy multiple biometric parameters are integrated into an identification system. Therefore, the primary goal of the multiple biometric grand challenge (MBCG), proposed in 2010, is to investigate, test and improve performance of face and iris recognition technology on both still and video imagery through a series of challenge problems and evaluation. The MBGC involves (1) face recognition on still unconstrained frontal, real-world-like high and low resolution imagery, (2) iris recognition from video sequences and off-angle images, (3) recognition from near infrared (NIR) and High Definition (HD) video streams and (4) fusion of face and iris (at score and image levels) for achieving total security. The challenge problems posed by MBGC may allow for fusion of face and iris at both the score level and the image level.

## 14.6   Focus of evaluation

Most of the evaluations available tend not to focus on some of the key problems that a system ultimately will need to deal with such as (1) large populations (the biometric double problem), (2) a significant age difference between gallery and probe image (the time delay or freshness/staleness problem) and (3) relatively uncontrolled environments (illumination, rotation, and background). It will be important for the development of FRT that technology evaluations incorporate more of these factors into the evaluation data set as design of the evaluation image set is fundamental to understanding the results achieved in the evaluation. The current evaluation methods do not necessarily include the evaluation of financial aspects as well as the evaluation of the ethical and political dimensions. More contextual and life cycle evaluations might be needed which might include financial evaluation as well as an ethical and political evaluation.

Taken together, the evaluations discussed above suggest that FRT has been proven effective for the verification task with relatively small populations in controlled environments. It seems that no single biometric will be able to do all the work (especially with regard to identification). As such, multi-biometric systems probably will be the future route of development. Evaluations may increasingly focus on multi-biometric systems in which face recognition systems may be one module.

# Conclusion

One of the reasons face detection and recognition has attracted so much research attention and sustained development over the past few decades is its great potential in numerous government and commercial applications. In 1997, at least 25 face recognition systems from 13 companies were available and since then, the numbers of face recognition systems and commercial enterprises have greatly increased owing to the emergence of many new application areas. Although some of these techniques are not publicly available for proprietary reasons, one can say that many others have been reported in research publications and available as affordable commercial systems.

The technology has evolved from laboratory research to many small, medium or large commercial deployments. At present, it is most promising for smaller medium-scale applications, such as office access control and computer log in; it still faces great technical challenges for large-scale deployments such as airport security and general surveillance.

Notwithstanding the extensive research effort that has gone into the area of face detection and recognition, a system that can be deployed effectively in an unconstrained setting is yet to evolve due to much variability in image parameters. Moreover, sensors and image capturing techniques also play a vital part in the overall success of a system. Another direction for improving recognition accuracy lies in a combination of multiple biometrics and security methods. It can work with other biometrics such as voice-based speaker identification, fingerprint recognition and iris scan in many applications.

The only system that does seem to work well in the face of these challenges is the human visual system. It makes sense, therefore, to attempt to understand the strategies this biological system employs, as a first step towards eventually translating them into machine-based algorithms. We believe that the future efforts may likely use such a path for refinement and performance improvement of present face recognition systems. Insights into the functioning of the human visual system may serve primarily as potentially fruitful starting points for computational investigations.

The issue of connectivity between the perception under the domain of cognitive science and the computer vision technique under the domain of computer technology is perhaps one of the interesting and current issue in face recognition. An interesting example is given in the associated figure to explain the statement. In this figure, the issue of perception will decide whether we are *seeing* the face image of Shakespeare or a landscape image consisting of trees

**FIGURE 14.6**: Visual perception and recognition illustrated

and cottages. Nothing has been said on this issue in this book. The central issue involved in this type of problem is related to mind-body interaction. The question arises regarding the process involved by which the brain - the material object of our body - can evoke correct signals in our mind, which in turn may control many of our actions or inactions for detecting or rejecting a face. This may inspire additional useful curiosity in intelligent face detection, which may go beyond the hard scientific or technological issues to the areas of reasoning and consciousness.

In the present book, the issues of realities of the mind-body problem have not been touched upon, but the realities which are under the realm of technology are discussed and investigated. An attempt has been made to realize the methodologies that can detect and recognize faces, even with some deviations in a limited sense of the practical world. Incidentally, using the human system for such detection and recognition is one aspect of coordination between visual perception, intelligence and muscle action which is related to the mind-body problems. However, one should be conscious and accept that the two kinds of realities are interdependent, which means that there can be a correlation between the two. Perhaps a day is not far off when these questions would be addressed in the language of science and translated in the domain of technology. Perhaps at the end, it may be quoted from the aphoristic book *Tractatus Logico-Philosophius* by L. Wittgenstein, *"what we cannot talk about we must pass over in silence."*

# Bibliography

[1] J. Wayman, A. K. Jain, D. Maltoni, and D. Maio. *Biometric Systems Technology, Design and Performance Evaluation.* Springer-Verlag London Limited, London, 2005.

[2] A. K. Jain, A. Ross, and S. Prabhakar. An introduction to biometric recognition. *IEEE Tran. on Circuits and Systems for Video Technology,* 14(1):20–26, 2004.

[3] T. Kanade. Picture procesing system by computer complex and recognition of human faces. *Ph.D thesis, Department of Information Science, Kyoto University,* November, 1973.

[4] M. Kirby and L. Sirovich. Application of the karhunen-loueve procedure for the characterization of human faces. *IEEE Trans. Pattern Analysis and Machine Intelligence,* 12(1):103–108, 1990.

[5] M. Turk and A. Pentland. Eigenfaces for recognition. *Journal of Cognitive Neurosciece,* 3:71–86, 1991.

[6] S. Li M. Tistarelli and R. Chellappa. *Handbook of Remote Biometrics for Surveillance and Security.* Springer, Berlin, 2009.

[7] P. Grother, P. J. Phillips, R. J. Micheals, D. M. Blackburn, E. Tabassi, and J. M. Bone. FRVT 2002: Evaluation report. *http://www.frvt.org/DLs/FRVT 2002 Evaluation Report.pdf,* 2003.

[8] P. J. Grother, G. W. Quinn, and P. J. Phillips. Report on the evaluation of 2d still image face recognition algorithms. *NIST Interagency report 7709, Multiple Biometric Evaluation(MBE),* June, 2010.

[9] A. Psarrou S. Gong, S. J. McKenna. *Dynamic Vision: From Images to Face Recognition.* Imperial College Press, World Scientific Publishing Company, UK, 2000.

[10] G. Chow and X. Li. Towards a system for automatic facial feature detection. *Pattern Recognition,* 26:17391755, 1993.

[11] H. Fronthaler K. Kollreider and J. Bigun. Evaluating liveness by face images and the structure tensor. *Fourth IEEE Workshop on Automatic Identification Advanced Technologies,* pages 75–80, 2005.

305

[12] R. W. Frischholz and A. Werner. Avoiding replay-attacks in a face recognition system using head-pose estimation. *IEEE International Workshop on Analysis and Modeling of Faces and Gestures*, pages 234–235, 2003.

[13] G. Chetty and M. Wagner. Multi-level liveness verification for face-voice biometric authentication. *Biometric Symposium*, Baltimore, USA, 2006.

[14] J. Li, Y. Wang, T. Tan, and A. Jain. Live face detection based on the analysis of fourier spectra. *Biometric Technology for Human Identification*, Proc. of SPIE, Baltimore, USA:296–303, 2004.

[15] P. Neuheisel K. Socolinsky, C. Wolff and B. Eveland. Illumination invariant face recognition using thermal infrared imagery. *Proc. of IEEE Conference on Computer Vision and Pattern Recognition*, Hawaii, 2001.

[16] L. E. Bahrick and R. Lickliter. Intersensory redundancy guides attentional selectivity and perceptual learning in infancy. *Developmental Psychology*, 36:190–201, 2000.

[17] L. E. Bahrick and R. Lickliter. Intersensory redundancyguides early perceptual and cognitive development. *Advances in child development and behavior*, 30:153–187, 2002.

[18] S. Carey. Becoming a face expert. *Philosophical Transactions of the Royal Society of London*, 335:95–103, 1992.

[19] S. Carey. *Perceptual classification and expertise in Perceptual and Cognitive Development*. Imperial College Press (World Scientific Publishing Company, Academic press, CA, USA, 1996.

[20] I. Leo C. Turati, E. Valenza and F. Simion. Three-month-olds visual preference for faces and its underlying visual processing mechanisms. *Journal of Experimental Child Psychology*, 90:255–273, 2005.

[21] E. McKone and B. L. Boyer. Sensitivity of 4-year-olds to featural and second order relational changes in face distinctiveness. *Journal of Experimental Child Psychology*, 94:134–162, 2006.

[22] C. A. Nelson. The development and neural bases of face recognition. *Infant and Child Development*, 10:3–18, 2001.

[23] R. Diamond and S. Carey. Why faces are not special: An effect of expertise. *Journal of Experimental Psychology: General*, 1986.

[24] T. Sakai, M. Nagao, and T. Kanade. Computer analysis and classification of photographs of human faces. In *First USAJapan Computer Conference*, 1972.

[25] I. Craw, H. Elli, and J. R. Lishman. Automatic extraction of face-feature. *Pattern Recognition Letters*, page 183187, 1987.

[26] R.Brunelli and T. Poggio. Face recognition: Feature versus template. *IEEE Trans. Pattern Analysis and Machine Intelligence*, 15(10):1042–1052, 1993.

[27] V. Govindaraju. Locating human faces in photographs. *International Journal of Computer Vision*, 19, 1996.

[28] X. Jiang, M. Binkert, B. Achermann, and H. Bunke. Towards detection of glasses in facial images. *Pattern Analysis and Applications*, 3:9–18, 2000.

[29] D. Marr and E. Hildreth. Theory of edge detection. In *Proc. of the Royal Society of London*, 1980.

[30] R. Herpers, H. Kattner, H. Rodax, and G. Sommer. An attentive processing strategy to detect and analyze the prominent facial regions. In *IEEE Proc. of Int. Workshop on Automatic Face and Gesture Recognition*, page 214220, 1995.

[31] K. M. Lam and H. Yan. Facial feature location and extraction for computerised human face recognition. In *Int. Symposium on information Theory and Its Applications*, 1994.

[32] R. Hoogenboom and M. Lew. Face detection using local maxima. In *IEEE Proc. of 2nd Int. Conf. on Automatic Face and Gesture Recognition*, 1996.

[33] A. F. Frangi J. Yang, D. Zhang and Yand. Two-dimensional pca: A new approach to face representation and recognition. *IEEE Transaction on Pattern Recognition and Machine Intelligence*, 26(1):131–137.

[34] X. G. Lv, J. Zhou, and C.-S. Zhang. A novel algorithm for rotated human face detection. In *IEEE Conf. on Computer Vision and Pattern Recognition*, 2000.

[35] C. Kotropoulos and I. Pitas. Rule-based face detection in frontal views. In *Proc. Int. Conf. on Acoustic, Speech and Signal Processing*, 1997.

[36] C. H. Lee, J. S. Kim, and K. H. Park. Automatic human face location in a complex background. *Pattern Recognition*, 29:18771889., 1996.

[37] G. Wei and I. K. Sethi. Face detection for image annotation. *Pattern Recognition Letters*, 20:13131321, 1999.

[38] J.C. Terrillon, M. Shirazi, H. Fukamachi, and S. Akamatsu. Comparative performance of different skin chrominance models and chrominance spaces for the automatic detection of human faces in color images. In

*Proc. Fourth IEEE International Conference on Automatic Face and Gesture Recognition*, 2000.

[39] N. Oliver, A. Pentland, and F. Berard. Lafter: A real-time face and lips tracker with facial expression recognition. *Pattern Recognition*, 33:13691382, 2000.

[40] J. Yang and A. Waibel. A real-time face tracker. In *IEEE Proc. of the 3rd Workshop on Applications of Computer Vision*, 1996.

[41] L. Yin and A. Basu. Integrating active face tracking with model based coding. *Pattern Recognition Letters*, 20:651–657, 1999.

[42] A. Witkin M. Kass and D. Terzopoulos. Snakes: active contour models. *Proc. of 1st Int Conf. on Computer Vision, London*, 1987.

[43] A. Lanitis, C. J. Taylor, and T. F. Cootes. Automatic tracking, coding and reconstruction of human faces using flexible appearance models. *IEEE Electronics Letters*, 30:15781579, 1994.

[44] A. L. Yuille, P. W. Hallinan, and D. S. Cohen. Feature extraction from faces using deformable templates. *Int. J. Comput. Vision*, 8:99111, 1992.

[45] A. Shackleton and W. J. Welsh. Classification of facial features for recognition. In *IEEE Proc. of Int. Conf. on Computer Vision and Pattern Recognition*, 1991.

[46] K. M. Lam and H. Yan. Locating and extracting the eye in human face images. *Pattern Recognition*, 29:771–779, 1996.

[47] C. Wongm, D. Kortenkamp, and M. Speich. A mobile robot that recognises people. In *IEEE Int. Conf. on Tools with Artificial Intelligence*, 1995.

[48] C. M. Liao. C. Han M. Y. Chern S. H. Jeng, H. Y and Y. T. Liu. Facial feature detection using geometrical face model: An efficient approach. *Pattern Recognition*, 31, 1998.

[49] O. Carmona F. Smeraldi and J. Bigun. Saccadic search with gabor features applied to eye detection and real-time head tracking. *Image Vision Comput*, 18:323329, 2000.

[50] Z. Liu and Y. Wang. Face detection and tracking in video using dynamic programming. In *Proc. International Conference Image Processing*, 2000.

[51] A. J. Colmenarez and T. S. Huang. Maximum likelihood face detection. In *IEEE Proc. of Second Int. Conf. on Automatic Face and Gesture Recognition*, 1996.

[52] A. J. Colmenarez, B. Frey, and T. S. Huang. Detection and tracking of faces and facial features. In *Proc. International Conference on Image Processing*, 1999.

[53] L. Sirovich and M. Meytlis. Symmetry, probability, and recognition in face space. *PNAS : Proceedings of the National Academy of Sciences*, 106(17):6895–6899, 2009.

[54] M. H. Yang. Kernel eigenfaces vs. kernel fisherfaces: Face recognition using kernel methods. *Proc. IEEE International Conference on Automatic Face and Gesture Recognition*, pages 215–220, 2002.

[55] A. Rosenfeld W. Zhao, R. Chellappa and P. J. Phillips. Face recognition: A literature survey. *ACM Computing Surveys*, pages 399–458, 2003.

[56] D. K. Baek and J. R. Beveridge. Pca vs. ica: A comparison on the feret data set. *Proc. of the Fourth International Conference on Computer Vision, Pattern Recognition and Image processing*, pages 824–827, 2002.

[57] H. Lu Q. Liu, R. Huang and S. Ma. Face recognition using kernel based fisher discriminant analysis. *Proc. of the fifth Int. Conf. on Automatic Face and Gesture Recognition, Washington DC*, 2002.

[58] S. T. Roweis and L. K. Saul. Nonlinear dimensionality reduction by locally linear embedding. *Science*, 290, 2000.

[59] G. Shakhnarovich and B. Moghaddam. Face recognition in subspaces. *Handbook of Face Recognition, Eds. S. Z. Li and A. K. Jain, Springer-Verlag*, page 35, 2004.

[60] X. Lu. Image analysis for face recognition. *Personal notes*, page 36, 2003.

[61] A. Martinez and A. Kak. Pca versus lda. *IEEE Trans. on Pattern Analysis and Machine Intelligence*, 23:228–233, 2001.

[62] P. Navarrete and J. R. D. Solar. Analysis and comparison of eigenspace-based face recognition approaches. *International Journal of Pattern Recognition and Artificial Intelligence*, 16:817–830, 2002.

[63] X. Wang and X. Tang. A unified framework for subspace face recognition. *IEEE Trans. on Pattern Analysis and Machine Intelligence*, 26, 2004, pages=1222-1228 ,.

[64] M. Grgic K. Delac and S. Grgic. Data independent comparative study of pca, ica, and lda on the feret set. *International Journal of Imaging Systems and Technology*, 15(5).

[65] B. Moghaddam A. Pentland and T. Starner. View-based and modular eigenspaces for face recognition. *IEEE Computer Conference on Computer Vision and Pattern Recognition*, pages 84–91, Jun. 1994.

[66] J. Hespanha P. Belhumeur and D. Kriegman. Eigenfaces vs. fisherfaces: Recognition using class specific linear projection. *Proc. of the Fourth European Conference on Computer Vision*, 1:45–58, 1996.

[67] M. S. Bartlett B. Draper, K. Baek and J. R. Beveridge. Recognizing faces with pca and ica. *Computer Vision and Image Understanding-Special Issue on Face Recognition*, 91(1):115–137, 2003.

[68] C. Liu and H. Wechsler. Comparative assessment of independent component analysis (ica) for face recognition. *Second International Conference on Audio- and Videobased Biometric Person Authentication*, March 1999.

[69] D. Cai X. He and P. Niyogi. Tensor subspace analysis. *Advances in Neural Information Processing Systems 18 (NIPS),*, 2005.

[70] K. Jung K. I. Kim and H. J. Kim. Face recognition using kernel principal component analysis. *Signal Processing Letters, IEEE ,,* pages 40–42, 2002.

[71] Y. Hu P. Niyogi X. He, S. Yan and H. J. Zhang. Face recognition using laplacian-faces. *IEEE Trans. Pattern Anal.Intell*, 2005.

[72] M. H. Yang. Face recognition using extended isomap. *Proc. of IEEE International Conference on Image Processing*, 2:117–120, 2002.

[73] T. M. Cover and P. E. Hart. Nearest neighbor pattern classification. *IEEE Trans. . Inform. Theory IT*, 13(1):21–27.

[74] R. O. Duda and P. E. Hart. *Pattern Classification and scene analysis.* John Wiley and sons, New York, 1973.

[75] M. A. Hearst, B. Schlkopf, S. Dumais, E. Osuna, and J. Platt. Trends and controversies - support vector machines. *IEEE Intelligent Systems*, 13(4):18–28, 1998.

[76] B. Scholkopf, A. Smola, and K. Muller. Nonlinaer component analysis as a kernel eigenvalue problem. *Neural Computation*, 10(5):1299–1319, 1998.

[77] D. Zhou and Z. Tang. Kernel-based improved discriminant analysis and its application to face recognition. *Soft computing A Fusion of Foundations, Methodologies and Applications*, 14:103–111, 2009.

[78] S. Tulyakov. Review of classifier combination methods. *Studies In Computational Intelligence (SCI)*, 90:361–386, 2008.

[79] A. J. Mansfield and J. L. Wayman. Best practices in testing and reporting performance of biometric devices,version 2.01. 2002.

[80] T. F. Cootes, C. J. Taylor, and A. Lanitis. Active shape models: Evaluation of a multi-resolution method for improving image search. *Proc. of British Machine Vision Conference*, pages 327–336, 1994.

[81] J. Colmenarez and T. S. Huang. Face detection with information-based maximum discrimination. *IEEE Proc. of International Conference on Computer Vision and Pattern Recognition*, 6:782–787, 1997.

[82] T. Strothotte and S. Schlechtweg. *Non-Photorealistic computer graphics. modeling, rendering, and animation*, volume CA, USA. Morgan Kaufmann Publishers Inc. San Francisco, 2002.

[83] B. Guo D. Terzopoulos Q. Zhang, Z. Liu and H. Y. Shum. Geometry driven photo realistic facial expression synthesis. *IEEE Trans. Visualization and computer Graphics*, 12(1):48–60, 2006.

[84] L. Wecker, F. Samavati, and M. Gavrilova. A multiresolution approach to iris synthesis. *Computer and Graphics*, 34(3):468–478, 2010.

[85] K. W. BowyerChang and P. Flynn. A survey of approaches and challenges in 3d and multi-modal 3d + 2d face recognition. *Computer Vision Image Understanding*, 101 (1):1–15, 2004.

[86] R. Lopez H. Tao and T. Huang. Tracking facial features using probabilistic network. *Third IEEE Internutional Conference on Automatic Face and Gesture Recognition*, (166, 1998.

[87] V. Blanz and T. Vetter. Face recognition based on fitting a 3d morphable model. *IEEE Tran. on Pattern Analysis and machine intelligence*, 25(9):1063–1074, 2003.

[88] R. Kumar, A. Barmpoutis, A. Banerjee, and B. C. Vemuri. Non-lambertian reflectance modeling and shape recovery of faces using tensor splines. *IEEE. Trans. on Pattern Analysis and Machine Intelligence*, 33(3):533–567, 2011.

[89] T. F. Cootes and C. J. Taylor. Active shape models-smart snakes. *Proc. of British Machine Vision Conference*, pages 266–275, 1992.

[90] H. Pyun, Y. Kim, W. Chae, H. W .Kang, and S. Y. Shin. An example-based approach for facial expression cloning. *Proceedings of ACM SIGGRAPH/Eurographics symposium on computer animation*, pages 167–176, 2003.

[91] F. H. Pighin, J. Hecker, D. Lischinski, R. Szeliski, and D. Salesin. Synthesizing realistic facial expressions from photographs. *SIGGRAPH*, pages 75–84, 1998.

[92] T. Beier and S. Neely. Feature-based image metamorphosis. *Computer Graphics*, 26(2):35–42, 1992.

[93] S. Romdhani, V. Blanz, C. Basso, and T. Vetter. Morphable models of faces. In A. Jain S. Li, editor, *Handbook of Face Recognition*, volume Springer-verlag, pages 217–245. 2005.

[94] H. Ip and L. Yin. Constructing 3D individualized head model from two orthogonal views. *The Visual Computer - the International Journal in Computer Graphics*, 12(5):254–266, 1996.

[95] I. T. Jolliffe. Principal component analysis. *Springer*, 1986. NY.

[96] Peter N. Belhumeur, P. Hespanha, and David J. Kriegman. Eigenfaces vs. fisherfaces: Recognition using class specific linear projection. *IEEE Trans. Pattern Analysis and Machine Intelligence*, pages 711–720, 1997.

[97] M. Bartlett, H. Ladies, and T. Sejnowski. Independent component representations for face recognition. *Proceedings of the SPIE: Conference on Human Vision and Electronic Imaging*, 3299:528–539, 2006.

[98] P. Comon. Independent component analysisa new concept? *Signal Process*, 36:287314, 1994.

[99] B.Moghaddam and A.Pentland. Probabilistic visual learning for object representation. *IEEE Trans. Pattern Anal. Mach. Intell.*, 19(7):696–710, 1997.

[100] Chengjun Liu. A bayesian discriminating features method for face detection. *IEEE Trans. Pattern Anal. Mach. Intell.*, 25(6):725–740, 2003.

[101] Peter Peer, Jure Kovac, and Franc Solina. Human skin colour clustering for face detection, 2003.

[102] Rein-Lien Hsu, Mohamed Abdel-mottaleb, and Anil K. Jain. Face detection in color images. *IEEE Transactions on Pattern Analysis and Machine Intelligence*, 24:696–706, 2002.

[103] Q. Zhu, S. Avidan, M. C. Yeh, and K. T. Cheng. Fast human detection using a cascade of histograms of oriented gradients. *Proceedings of Computer Vision and Pattern Recognition*, 2:1491–1498, 2006.

[104] V. M. Bennine I. L. Thomas and N. P. Ching. *Classification of Remotely Sensed Images*. (Bristol: Hiller), 1987.

[105] W. L. G. Knootz, P. M. Narendra, and K. Fukunaga. A graph-theoretic approach to non-parametric clustering. *IEEE Trans. on Computers*, C-25:936–944, 1976.

[106] B. Kartikeyan, A. Sarkar, and K. L. Majumder. A segmentation approach to classification of remote sensing imagery. *International Journal of Remote Sensing*, 19(9):1695–1709, 1998.

[107] A. Khotanzad and A. Bouarfa. Image segmentation by a parallel, non-parametric histogram based clustering algorithm. *Pattern Recognition*, 23:961–944, 1990.

[108] C. Zhang and Z. Zhang. Winner-take-all multiple category boosting for multi-view face detection. *Technical report, Microsoft Research*, MSR-TR-2009-190, 2009.

[109] A. M. Martinez and A. C. Kak. Pca versus lda. *IEEE Trans. Pattern and Machine Intelligence*, 23(2):228–233, 2001.

[110] H.A.Rowley, S.Baluja, and T.Kanade. Neural network based face detection. *IEEE Trans. Pattern Analysis and Machine Intelligence*, 20(1):23–38, 1998.

[111] K. K. Sung and T. Poggio. Example-based learning for view-based human face detection. *IEEE Trans.Pattern Anal. Mach. Intelligence*, 20,:39–51, 1998.

[112] P. Viola and M. Jones. Rapid object detection using a boosted cascade of simple features. In *Proc. Conf. Computer Vision and Pattern Recognition*, pages 511–518, 2001.

[113] P. Viola and M. Jones. Robust real-time face detection. *International Journal of Computer Vision*, 57(2), 2004.

[114] N. Ahuja M.-H. Yang and D. Kriegman. Face detection using mixtures of linear subspaces. *Proceedings Fourth IEEE International Conference on Automatic Face and Gesture Recognition*, 2000.

[115] D. G Kirkpatrickand H. Edelsbrunner and R.Seidel. On the shape of a set of points in a plane. *IEEE Trans. On Inform. Theory*, vol. IT-29, year=1983, pages=551-559 ,.

[116] J. Geffroy. Surun problme destimation gomtrique. *Publications de l'Institut de Statistique de l'Universit de Paris*, 13:191–210, 1964.

[117] A. Renyl and R. Sulanke. ber die konvexe hlle von n zufllig gcwhlten punktenz. wahrscheinlichkeitstheorie und verw. gebiete. 2:75–84, 1963.

[118] L. Devroye and G. L. Wise. Detection of abnormal behavior via nonparametric estimation of the support. *SIAM Journal of Applied Mathematics*, 38(3):480–488, 1980.

[119] A. Rodriguez-Casal. Set estimation under convexity type assumptions. *Annales de IInstitut Henri Poincar- Probability and Statistics*, 43:763–774, 2007.

[120] C. A. Murthy D. P. Mandal and S. K. Pal. Determining the shape of a pattern class from sampled points: Extension to rn. *Int. J. of General Systems*, vol. 26, no. 4:293–320, 1997.

[121] D. P. Mandal and C. A. Murthy. Selection of alpha for alpha-hull in r2. *Pattern Recognition*, 30, (10):1759–1767, 1997.

[122] U. Grenander. Abstract inference. 1981.

[123] H. Kestelman. Lebesgue measure. *Ch. 3 in Modern Theories of Integration*, 2nd rev. ed. New York: Dover:67–91, 1960.

[124] C. A. Murthy. *On Consistent Estimation Of Classes in The Context of Cluster Analysis*. PhD thesis, Indian Statistical Institute, India, 1988.

[125] S. A. Nene, S. K. Nayar, and H. Murase. COIL database. *Technical Report* CUCS-*006-96*, 1996.

[126] OLIVETTI face database. *http://www.cam-orl.co.uk/facedatabase.html*.

[127] *http://cvc.yale.edu/projects/yalefaces/yalefaces.html*.

[128] A.R. Martinez and R. Benavente. The ar face database. Technical report, Computer Vision Center (CVC), Barcelona, 1998.

[129] J. Phillips, H Moon, S Rizvi, and P Rauss. The feret evaluation methodology for face-recognition algorithms. *IEEE Trans. on Pattern Analysis and Machine Intelligence*, 22(10), 2000.

[130] International Biometrics Association(IBA):. Standard biometric industry terminology and definitions: Draft. *IBA Standard Washington, DC*, BSC 2.61987R, 1987.

[131] Chengjun Liu and Harry Wechsler. Evolutionary pursuit and its application to face recognition. *IEEE Trans. Pattern Anal. Mach. Intell.*, 22(6):570–582, 2000.

[132] B. Kumar. Tutorial survey of composite filter designs for optical correlators. *Applied Optics*, 31(23):4773–4801, 1992.

[133] A.B.VaderLugt. Signal detection by complex matched spatial filtering. *IEE Transaction on Information Theory*, IT-10:139–145, 1964.

[134] B. Kumar, M. Savvides, C. Xie, K. Venkataramani, J. Thornton, and A.Mahalanobis. Biometric verification with correlation filters. *Applied Optics*, 43(2):391–402, 2004.

[135] C. Hester and D. Casasent. Multivariate technique for multiclass pattern recognition. *Applied Optics*, 9(11), 1980.

[136] B. Kumar. Minimum variance synthetic discriminant functions. *Journal of Optical Society of America,A*, 3(10):1579–1584, 1986.

[137] A. Mahalanobis, B. Kumar, and D. Casassent. Minimum average correlation energy filter. *Applied Optics*, 26(17):3633–3640, 1987.

[138] Ph.Refregier. Filter design for optical pattern recognition: Multi-criteria optimization approach. *Optics Letters*, 15:854–856, 1990.

[139] G.Ravichandran and D.Casasent. Minimum noise and correlation energy optical corrrelation filter. *Applied Optics*, 31:1823–1833, 1992.

[140] A. Mahalanobis, B. Kumar, S. Song, S. Sims, and J. Epperson. Unconstrained correlation filter. *Applied Optics*, 33:3751–3759, 1994.

[141] M. Savvides, B. Kumar, and P.K. Khosla. Two-class minimax distance transform correlation filter. *Applied Optics*, 41:6829–6840, 2002.

[142] B. Kumar, D. Carlson, and A. Mahalanobis. Optimal trade-off synthetic discriminant function filters for arbitrary devices. *Optics Letters*, 19(19):1556–1558, 1994.

[143] Ph. Refregier. Optimal trade-off filters for noise robustness, sharpness of the correlation peak, and horner efficiency. *Optics Letters*, 16(11):829–831, 1991.

[144] M. Alkanhal, B. Kumar, and A. Mahalanobis. Improving the false alarm capabilities of the maximum average correlation height correlation filter. *Optical Engineering*, 39(5):1133–1141, 2000.

[145] B. Kumar and M. Alkanhal. Eigen-extended maximum average correlation height (eemach) filters for automatic target recognition. *Proc. SPIE 4379, Automatic Target Recognition XI*, 4379:424–431, 2001.

[146] R. Muise, A. Mahalanobis, R. Mohapatra, X. Li, D. Han, and W. Mikhael. Constrained quadratic correlation filters for target detection. *Applied Optics*, 43(2):304–314, 2004.

[147] A. Mahalanobis, R. Muise, S. R. Stanfill, and A. V. Nevel. Design and application of quadratic correlation filters for target detection. *IEEE Trans. on Aerospace Electronic Systems*, 40(3):837–850, 2004.

[148] N. V. Alan and A. Mahalanobis. Comparative study of maximum average correlation height filter variants using ladar imagery. *Optical Engineering*, 42(2):541–550, 2003.

[149] S. Goyale, N. K. Nischal, V. K. Beri, and A. K. Gupta. Wavelet-modified maximum average correlation height filter for rotation invariance that uses chirp encoding in a hybrid digital–optical correlator. *Applied Optics*, 45(20):4850–4857, 2006.

[150] A. Aran, N. K. Nishchal, V. K. Beri, and A. K. Gupta. Log-polar transform-based wavelet-modified maximum average correlation height filter for distortion-invariant target recognition. *Optics and Lasers in Engineering*, 46(1):34–41, 2008.

[151] H. Lai, V. Ramanathan, and H. Wechsler. Reliable face recognition using adaptive and robust correlation filters. *Computer Vision and Image Understanding*, 111:329–350, 2008.

[152] K. H. Jeong, W. Liu, S. Han, E. Hasanbelliu, and J.C. Principe. The correntropy mace filter. *Pattern Recognition*, 42(9):871–885, 2009.

[153] M. D. Rodriguez, J. Ahmed, and M. Shah. Action mach a spatio-temporal maximum average correlation height filter for action recognition. In *CVPR*. IEEE Computer Society, 2008.

[154] D. Bolme, B. Draper, and R. Beveridge. Average of synthetic exact filters. In *CVPR*, pages 2105–2112. IEEE, 2009.

[155] D. Bolme, J. Beveridge, B. Draper, and Y. M. Lui. Visual object tracking using adaptive correlation filters. In *CVPR*, pages 2544–2550. IEEE, 2010.

[156] A. Rodriguez, V. N. Boddeti, B. Kumar, and A. Mahalanobis. Maximum margin correlation filter: A new approach for localization and classification. *IEEE Trans. on Image Processing*, 22(2):631–643, 2012.

[157] J. C. Platt. Advances in kernel methods. chapter Fast training of support vector machines using sequential minimal optimization, pages 185–208. MIT Press, 1999.

[158] K. A. Mashouq, B. Kumar, and M. Alkanhal. Analysis of signal-to-noise ratio of polynomial correlation filters. In *Proc. SPIE*, pages 407–413, 1999.

[159] A. Mahalanobis and B. Kumar. Polynomial filters for higher-orger and multi-input information fusion. In *11th Euro-Am. Opto-Electronic Information Processing Workshop*, pages 221–231, 1999.

[160] D. Casasent. Unified synthetic discriminant function computational formulation. *Applied Optics*, 23(10):1620–1627, 1984.

[161] B. Kumar, A. Mahalanobis, and R. Juday. *Correlation Pattern Recognition*. Cambridge University Press, 2005.

[162] M. Savvides and B. Kumar. Quad-phase minimum average correlation energy filters for reduced-memory illumination-tolerant face authentication. In *Audio and Visual Biometrics based Person Authentication*, pages 19–26, 2003.

[163] M. Savvides, B. Kumar, and P. Khosla. Robust, shift-invariant biometric identification from partial face images. In *Biometric Technology for Human Identification. Ed. A. K. Jain and N. K. Ratha Anil*, volume Proc. SPIE. 5404, pages 124–135, 2004.

[164] M. Savvides C. K. Ng and P. K. Khosla. Real time face verification on a cell phone using advanced correlationfilters. In *Proc. of IEEE Workshop on Automatic Identification Advanced Technologies*, volume 57, 2005.

[165] R. Abiantun C. Xie J. Heo, M. Savvides and B. Kumar. Face recognition with kernel correlation filters on a large scale database. In *Proc. IEEE Int. Conf. on Acoustics,Speech and Signal Processing*, volume II, page 181, 2006.

[166] M. Savvides and B. Kumar. Quad-phase minimum average correlation energy filters for reduced-memory illumination-tolerant face authentication. In *Audio and Visual Biometrics based Person Authentication*, 2003.

[167] A. Alfalou Y. Ouerhani, M. Jridi and C. Brosseau. Optimized pre-processing input plane gpu implementation of an optical face recognition technique using a segmented phase only composite filter. *Optics Communications*, 289:33–44, 2013.

[168] S. R. Everardo, J. A. Gonzalez-Fraga, and J. I. Ascencio-Lopez. Performance of correlation filters in facial recognition. In *Pattern Recognition*, volume 6718 of *Lecture Notes in Computer Science*, pages 155–163. Springer Berlin Heidelberg, 2011.

[169] M. Savvides, B. Kumar, and P. Khosla. Robust, shift-invariant biometric identification from partial face images. In *Biometric Technology for Human Identification. Edited by A. K. Jain and N. K. Ratha*, volume Proc. SPIE 5404, pages 124–135, 2004.

[170] M. Savvides, J. Heo, R. Abiantun, C. Xie, and B. Kumar. Partial and holistic face recognition on frgc-ii data using supportvector machines. In *Proc. of IEEE ICCVPR*, volume 48, 2006.

[171] M. Savvides and B. Kumar. Illumination normalization using logarithm transforms for face authentication. In *Proc. of Int. Conf. on Advances in Pattern Recognition, Springer*, 2003.

[172] J. Heo, M. Savvides, and B. Kumar. Performance evaluation of face recognition using visual and thermal imagery with advanced correlation filters. In *Proc. of IEEE ICCVPR*, volume 9, 2005.

[173] B. Kumar, M.Savvides, K. Venkataramani, and C. Xie. Spatial frequency domain image processing for biometric recognition. In *IEEE ICIP*, 2002.

[174] Advanced multimedia processing lab web page at electrical and computer engineering department at cmu, http://amp.ece.cmu.edu.

[175] M. Savvides, B. Kumar, and P. Khosla. Face verification using correlation filters. 2002.

[176] M. Savvides and B. Kumar. Efficient design of advanced correlation filters for robust distortion-tolarant face recognition. In *Proc. of the IEEE Conference on Advanced Video and Signal Based Surveillance*, 2003.

[177] M. Savvides, Venkataramani, and B. Kumar. Incremental updating of advanced correlation filters for biometric authentication systems. In *Proc. of IEEE Int. Conf. on Multimedia and Expo*, volume III, page 229, 2003.

[178] M. Savvides and B. Kumar. Illumination normalization using logarithm transformsfor face authentication. *Lecture Notes in Computer Science, Springer*, 2688:549–556, 2003.

[179] M. Savvides, B. Kumar, and P.K. Khosla. Cancellable biometric filters for face recognition. In *International Conference in Pattern Recognition (ICPR)*, volume 3, pages 922–925, 2004.

[180] M. Savvides, B. Kumar, and P.K. Khosla. Corefaces- robust shift invariant pca based correlation filter forillumination tolerant face recognition. In *Proc. of the 2004 IEEE Computer Society Conference on Computer Vision and Pattern Recognition (CVPR04)*, 2004.

[181] J. Thornton, M. Savvides, and B. Kumar. Linear shift-invariant maximum margin svm correlation filter. In *IEEE ISSNIP*, 2004.

[182] R.Patnaik and D.Casasent. Illumination invariant face recognition and impostor rejection using different minace filter algorithms. In *Proc. SPIE 5816, Optical Pattern Recognition XVI*, volume 94, pages 94–104, 2005.

[183] C. Xie, M. Savvides, and B. Kumar. Quaternion correlation filter for face recognition in wavelet domain. In *IEEE ICASSP*, 2005.

[184] C. Xie and B. Kumar. Face class code based feature extraction for face recognition. In *Fourth IEEE Workshop on Automatic Identification Advanced Technologies*, 2005.

[185] C. Xie and B. Kumar. Comparison of kernel class-dependence feature analysis (kcfa) with kernel discriminant analysis (kda) for face recognition. In *IEEE*, 2007.

[186] C. Xie, M. Savvides, and B. Kumar. Kernel correlation filter based redundant class-dependence feature analysis (kcfa) on frgc2.0 data. *LNCS Springer*, 3723:32–43, 2005.

[187] R. Abiantun, M. Savvides, and B. Kumar. Generalized low dimensional feature subspace for robust face recognition on unseen datasets using kernel correlation feature analysis. In *IEEE ICASSP*, 2007.

[188] Yan Yan and Yu-Jin Zhang. Tensor correlation filter based class-dependence feature analysis for face recognition. *Neurocomputing*, 71(1618):3434 – 3438, 2008.

[189] Y. Yan and Y. Zhang. 1d correlation filter based class-dependence feature analysis forface recognition. *Pattern Recognition*, 41:3834–3841, 2008.

[190] M. Levine and Y. Yu. Face recognition subject to variations in facial expression, illumination and pose using correlation filters. *Computer Vision and Image Understanding*, 104 (2006) 115:1–15, 2006.

[191] Mohammadreza Maddah and Saeed Mozaffari. Face verification using local binary pattern-unconstrained minimum average correlation energy correlation filters. *J. Opt. Soc. Am. A*, 29(8):1717–1721, 2012.

[192] R. Rodrguez, H. Mndez-Vzquez, and E. Garca-Reyes. Illumination invariant face recognition using quaternion-based correlation filters. *Journal of Mathematical Imaging and Vision*, 45 (2):164–175, 2013.

[193] D. W .Carlson A. Mahalanobis and B. Kumar. Evaluation of mach and dccf correlation filters for sar atr using the mstar public database. volume 3370, pages 460–468, 1998.

[194] S. Bhuiyan, M. A .Alam, S. Mohammad, S. Sims, and F. Richard. Target detection, classification, and tracking using a maximum average correlation height and polynomial distance classification correlation filter combination. *Optical Engineering*, 45(11):116401–13, 2006.

[195] P. K. Banerjee and A. K. Datta. Generalized regression neural network trained preprocessing of frequency domain correlation filter for improved face recognition and its optical implementation. *Optics and Laser Technology*, 45:217 – 227, 2013.

[196] B. Kumar A. Mahalanobis and S. Sims. Distance classifier correlation filters for multiclass target recognition. *Applied Optics*, 35(17)(17):3127–3133, 1996.

[197] C. Sanderson. *Biometric Person Recognition: Face, Speech and Fusion.* VDM-Verlag, 2008.

[198] G. Catalano, A. Gallace, B.Kim, S. Pedro, and F.Santoro. Optical flow. Technical report, 2009.

[199] M. Savvides, K. Venkataramani, and B. Kumar. Incremental updating of advanced correlation filters for biometric authentication systems. In *Proc. of IEEE Int. Conf. on Multimedia and Expo*, volume III, page 229, 2003.

[200] M. Savvides and B. Kumar. Quad phase minimum average correlation energy filters for reduced memory illumination tolerant face authentication. In *Proc. of Int. Conf. on Advances in Pattern Recognition*, 2003.

[201] Martine David Levine and Yingfeng Yu. Face recognition subject to variations in facial expression, illumination and pose using correlation filters. *Computer Vision and Image Understanding*, 104:1–15, 2006.

[202] B. Kumar and E. Pochapsky. Signal-to-noise ratio considerations in modified matched spatial filters. *J. Opt. Soc. Am. A*, 3:777–786, 1986.

[203] X. Liu, T. Chen, and B. Kumar. Face authentication for multiple subjects using eigenflow. *Pattern Recognition, Special issue on Biometric*, 36:313–328, 2003.

[204] X. Jiang, B. Mandal, and A. C. Kot. Eigenfeature regularization and extraction in face recognition. *IEEE Trans. on Pattern Analysis Machine Intelligence*, 30(3):383–394, 2008.

[205] A.Mahalanobis and B. Kumar. Polynomial filters for higher order correlation and multi-input information fusion. Patent, US6295373 B1, 1997.

[206] P. K. Banerjee and A. K. Datta. Class specific subspace dependent nonlinear correlation filtering for illumination tolerant face recognition. *Pattern Recognition Letters*, 36:177 – 185, 2014.

[207] K.C. Lee, J. Ho, and D. Kriegman. Acquiring linear subspaces for face recognition under variable lighting. *IEEE Trans. on Pattern Analysis and Machine Intelligence*, 27(5):684–698, 2005.

[208] T. Sim, S. Baker, and M. Bsat. The cmu pose, illumination, and expression database,. *IEEE Transactions on Pattern Analysis and Machine Intelligence*, 25(12):1615–1618, 2003.

[209] P. K. Banerjee and A. K. Datta. Illumination and noise tolerant face recognition based on eigen-phase correlation filter modified by mexican hat wavelet. *Journal of Optics*, 38(3):160–168, 2009.

[210] J. Huang and H. Wechsler. Visual routines for eye location using learning and evolution. *IEEE Trans. on Evolutionary Computation*, 1999.

[211] S. Welstead. Fractal and wavelet image compression techniques. *SPIE Press*, 1999.

[212] Elzbieta Marszalec, Birgitta Martinkauppi, Maricor Soriano, and Matti Pietikainen. Physics-based face database for color research. *Journal of Electronic Imaging*, 9(1):32–38, 2000.

[213] *http://www.ee.surrey.ac.uk/CVSSP/xm2vtsdb/*.

[214] *Face Recognition: From Theory to Applications,*. NATO ASI Series F, Computer and Systems Sciences,Springer, 1998.

[215] D. Blackburn, M. Bone, and P. J. Phillips. Face recognition vendor test 2000. *Technical report*, page http://www.frvt.org, 2001.

[216] P. J. Phillips, P. Grother, R. J. Micheals, D. M. Blackburn, E. Tabassi, and M. Bone. Face recognition vendor test 2002, evaluation report. pages 1–56, 2003.

[217] P. J. Phillips, W. T. Scruggs, A. J. O'Toole, P. J. Flynn, K. W. Bowyer, C. L. Schott, and M. Sharpe. Frvt 2006 and ice 2006 large-scale results. *National Institute of Standards and Technology*, pages 1–56, 2007.

[218] J. R. Beveridge, G. H. Givens, P. J. Phillips, B. A. Draper, D. S. Bolme, and Y. M. Lui. Quo vadis face quality. *Image and vision Computing*, 28, 2010.

# Index

(RGB) color space, 85
2DPCA, 46

Ability to verify rate, 15
ACE measure, 176
Active learning, 116
Active shape model, 25
AdaBoost, 127
artificial neural networks, 107
ASM measure, 176
Automated face recognition , 5
average correlation height, 267
Average exact synthetic filter, 261

Backpropagation, 110
Backpropagation algorithm, 110
Bayes classifier, 92
Bayes rule, 70
Bayesian decision boundary, 70
Bayesian decision theory, 63
Bayesian discriminating feature,
      75
bi-class classification, 137
Bimodal audiovisual, 16
Biometric system, 3
Biometrics, 1
Bivariate histogram, 99

Cascade classifier, 131
Chromosome in GA, 167
Class scatter, 53
Class specific subspace, 226
Closed test, 9
Clusters, 100
CMU-PIE dataset, 295
Cognitive psychology, 16

Color, 83
Color model, 84
color space, 84
Colour information, 23
Configural processing, 16
Constrained correlation filters, 175
Correlation based method, 32
Correlation classifiers, 244
correlation filters, 172
Correlation peak, 174

d-prime value, 15
DCCF filter, 212
Decision cost function, 15
Detection in infrared, 97

ECP-SDF filter, 175
Edges, 22
Eigenface, 44
eigenface subspace, 226
Eigenvalues, 43
Eigenvectors, 43
Elastic bunch graph, 253
Ellipse fitting, 269
EMACH filter, 177
Equal error rate (EER), 15
Euclidean distance, 139
Evolutionary pursuit, 161, 168
Eye detection, 263
Eye mapping, 93

Face bunch graph, 260
Face direction, 118
Face generation, 280
Face graph, 260
face landmarks, 259